Parija's Immunologic Diagnostics in Parasitology

The field of parasitology has seen tremendous progress in the last two to three decades with the advent of molecular biology and its application to the study of parasites. This has opened up new vistas into the understanding of the taxonomy, pathogenesis, and immunological mechanisms of many parasites, along with the design of improved diagnostic tests and application of molecular epidemiology in tracking and controlling these diseases. *Parija's Immunologic Diagnostics in Parasitology: Principles, Techniques, and Applications* is an attempt to cover the existing information along with newer findings so that up-to-date information can be condensed into one volume.

The book has been divided into three parts. Part I deals with the background of immunological techniques including history, principles, sample collection, and specimen processing, as well as quality assurance of these tests. Part II covers detailed discussion on the important existing immunological test formats along with step-by-step descriptions for development of experimental or in-house tests. Part III includes chapters on recent tests gaining importance like point-of-care tests and immuno-PCR. The book includes four appendices covering design of a serological laboratory, commercial suppliers of immunodiagnostic kits for parasitic diseases, a list of suppliers for biologicals, chemicals, and labware, and the Indian Academy of Tropical Parasitology guidelines for parasitic disease diagnoses.

Parija's Immunologic Diagnostics in Parasitology: Principles, Techniques, and Applications is an effort to inform and engage a wide spectrum of audience members, including postgraduate or doctoral students and faculty members across the medical and veterinary sciences disciplines. The text is intended both as a textbook and a procedure handbook.

Diagnostic Parasitology

Series Editor: *Subhash Chandra Parija*

This book series aims to serve as a comprehensive manual on various diagnostic methods in human parasitology, covering both common and specialized techniques used for the diagnosis of parasitic diseases of human, veterinary, and environmental studies. It encompasses various diagnostic strategies, such as microscopy, immunology, molecular diagnostics, and cutting-edge applications of artificial intelligence. This series intends to provide detailed laboratory diagnostic methods as part of a one-health approach to parasitic infections, offering valuable insights for professionals and researchers in the field.

Parija's Immunologic Diagnostics in Parasitology: Principles, Techniques, and Applications

Subhash Chandra Parija, Abhijit Chaudhury

Parija's Immunologic Diagnostics in Parasitology

Principles, Techniques, and Applications

Subhash Chandra Parija
Abhijit Chaudhury

CRC Press
Taylor & Francis Group
Boca Raton London New York

CRC Press is an imprint of the
Taylor & Francis Group, an **informa** business

Designed cover image: Shutterstock

First edition published 2027
by CRC Press
2385 NW Executive Center Drive, Suite 320, Boca Raton FL 33431

and by CRC Press
4 Park Square, Milton Park, Abingdon, Oxon, OX14 4RN

CRC Press is an imprint of Taylor & Francis Group, LLC

ISBN: 978-1-041-08199-9 (hbk)
ISBN: 978-1-041-08200-2 (pbk)
ISBN: 978-1-003-64422-4 (ebk)

DOI: 10.1201/9781003644224

Typeset in Times
by SPi Technologies India Pvt Ltd (Straive)

Dedicated to

The Supreme Power who has willed and given us the strength to take up this work.

To my brother, the late Shri Kailash Chandra Parija, and to my sister-in-law, the late Smt Satyabhama Parija, for their blessings.

To my wife, Ms. Jyotirmayee Parija for her selfless support.

And to my professional colleagues and mentors for their guidance.

Subhash Chandra Parija

To my wife Oiendrilla and my children Madhulika and Shashank for their support and encouragement.

And to my mother, the late Smt Manjari Chaudhury, and father, the late Dr. Prasanta Kumar Chaudhury, for their blessings.

Abhijit Chaudhury

Contents

UNIT 2 Miscellaneous Tests

PART III Recent Trends in Immunodiagnostics of Parasitic Infections

About the Authors

Subhash Chandra Parija is the former vice-chancellor of Shri Balaji Vidyapeeth, Pondicherry. He is the former director and dean (research) of the Jawaharlal Institute of Postgraduate Medical Education & Research (JIPMER), Pondicherry. He obtained an MBBS in 1977 from Utkal University, Cuttack; an MD in microbiology in 1981 from Banaras Hindu University (BHU), Varanasi; and a PhD in microbiology in 1987 from University of Madras, Chennai. He is one of the few medical microbiologists in India to be conferred a DSc in microbiology, the highest degree in research, for his contribution to the field of medical parasitology by the University of Madras, Chennai. Prof. Parija was honored with the Distinguished BHU Alumnus Award in the field of medical sciences during the 150th anniversary celebration of BHU in 2012. Every year since 2020 Prof. Parija has been listed among the Top 2% of Scientists in the World by Stanford University. He has nearly four decades of teaching and research experience, contributing mainly to infectious diseases, especially parasitology, diagnostics, and public health. He is actively involved in conducting research programs on immunology, epidemiology, and simple, cost-effective diagnostic tests for parasitic diseases for communities with less resources. Prof. Parija and his team have greatly contributed to clinical and diagnostic parasitology. They were the first to demonstrate and detect excreted parasitic antigens in urine for diagnosis of hydatid disease and cysticercosis; report detection of *Entamoeba histolytica* DNA in the urine and saliva for diagnosis of amoebic liver abscess; document *Entamoeba moshkovskii* from India; evaluate multiplex PCR for simultaneous detection of *Entamoeba histolytica*, *Entamoeba dispar*, and *Entamoeba moshkovskii* in stool specimens for diagnosis of intestinal amoebiasis; and document the subtype identification of *Blastocystis* in Pondicherry. Prof. Parija created innovations in stool microscopy techniques by using LPCB, KOH, wet mount preparation, thick smear, and formalin-acetone methods for better identification of parasitic diseases.

Prof. Parija is the author of more than 450 research papers and has supervised more than 85 PhDs, MDs, and other postgraduates. He has two published patents, 37 copyrights, and one technology transfer. Prof. Parija authored and/or edited 19 books including the popular *Textbook of Parasitology* and, most recently, *Effective Medical Communication: The A, B, C, D, E of It* (Springer). He is a fellow of the Royal College of Pathologists, London and the International Academy of Medical Sciences, New Delhi. He is also a fellow of many professional bodies of eminence such as the National Academy of Medical Sciences, New Delhi; the Indian College of Pathologists, New Delhi; the Indian Academy of Tropical Parasitology, Pondicherry; and many others. Prof. Parija is a member of the Expert Committee of the Food and Agricultural Organization – World Health Organization (FAO-WHO) for formulating guidelines on food safety for parasites. He chaired the Committee for Pondicherry Declaration on the Identification and Detection of *Entamoeba histolytica* and transferred the cost-effective diagnostic tests in parasitic diseases to the University of Parendenya, Sri Lanka under the aegis of the Indo-Sri Lanka Joint program on technology transfer. He has served on the academic, examination, and research advisory committees of many national and international institutes including the BP Koirala Institute of Health Sciences, Dharan, Nepal; College of Medicine & Health Sciences, Colombo University, Sri Lanka; Sultan Quaboos University, Muscat, Oman; and the Faculty of Medicine, University of Malaya, Malaysia. He is currently the editor-in-chief of *Tropical Parasitology*, in addition to being a member of the editorial board of many peer-reviewed journals.

Prof. Parija founded the Indian Academy of Tropical Parasitology, the only such organization in India, started *Tropical Parasitology*, a scientific journal in parasitic diseases, initiated the IATP-quality assurance program for parasitic diseases in India, and began the postdoctoral program in tropical parasitology currently being introduced in many premier medical institutes across the country. He also founded the Health & Intellectual Property Rights Academy (HIPRA), a unique body promoting the knowledge, study, and health care intellectual property rights of India. He has received more than 29 awards, both international and national, including BPKIHS Internal Oration Award (1997) of the B.P. Koirala Institute of Health Sciences, Nepal; Dr. B.C. Roy National Award (2003) of the Medical Council of India; Dr. R.V. Rajam Oration Award (2019) and Dr. P.N. Chuttani Oration Award (2007) of the National Academy of Medical Sciences; Prof. B.K. Aikat Oration Award (1998) and Major General Saheb Singh Sokhey Award (1992) of the Indian Council of Medical Research; Dr. Subramaniam Memorial Oration Award (2015) of the Indian Association of Biomedical Scientists; Sri S.M. Ismail Oration Award (2005) of the Indian Association for Development of Veterinary Parasitology; Dr. B.P. Pandey Memorial Oration Award (1998) of the Indian Society for Parasitology; Dr. S.C. Agarwal Oration Award (2001) of the Indian Association of Medical Microbiologists; and Dr. B.P. Pandey Memorial Oration Award (1998) of the Indian Society for parasitology, among others.

Prof. Parija has contributed immensely toward institutional building and medical education and has mentored undergraduates, postgraduates, PhD scholars, and young faculty to use their interests in research and carriers in parasitic diseases of public health importance.

Abhijit Chaudhury, MBBS (Calcutta University), MD (BHU, Varanasi), DNB, D (ABMM), is a professor in the Department of Microbiology at Sri Venkateswara Institute of Medical Sciences and its affiliated Sri Padmavathy Medical College for Women at Tirupati, Andhra Pradesh, India. He started his career in microbiology at the Himalayan Institute of Medical Sciences, Dehradun and then moved to the Manipal College of Medical Sciences, Pokhara, Nepal; and since 1998, he has been associated with the present institutions. He also taught as a faculty member of microbiology at the Medical University of St. Eustatius, Netherland Antilles; SSR Medical College, Mauritius; and Oman Medical College.

Dr. Chaudhury's research interests include pathogenesis of protozoan parasites, mycobacterial genomics, and biofilm infections. He was a member of the Task Force for Curriculum Development for Laboratory Technology courses by the Ministry of Health and Family Welfare, Government of India, and a reviewer for *Antibiotic Treatment Guidelines* published by the government of West Bengal. He has served as referee for several national and international journals including *Indian Journal of Medical Research*, *Frontiers in Microbiology*, *Microbial Pathogenesis*, and *FEMS Immunology and Medical Microbiology*. He has more than 25 years of teaching experience in medical microbiology for undergraduate and postgraduate students and has supervised or co-supervised the MSc, MD, and PhD work of several students. He has been instrumental in designing horizontally and vertically integrated teaching modules for the MBBS students years before the introduction of competency-based medical education by the Medical Council of India. He has also published more than 100 research articles in peer-reviewed national and international journals, has been the coeditor of a book on parasitic zoonosis, and has authored seven book chapters. Dr. Chaudhury is a member of many scientific societies including the Indian Association of Medical Microbiology and the Indian Academy of Tropical Parasitology, and he is an assessor for the National Medical Council (formerly Medical Council of India) and National Assessment and Accreditation Council, India. He is the former chief managing editor of *Tropical Parasitology*, the official publication of the Indian Academy of Tropical Parasitology. He has been the recipient of the prestigious fellowship of the Indian Academy of Tropical Parasitology and the Dr. A.N. Chakrabarti Memorial Award of the Indian Association of Medical Microbiologists for the best published paper on mycobacteriology.

Preface

The world of parasites is astoundingly huge. The earth is the habitat for 65,000 protozoan and more than 75,000 helminth species, but the great majority of them are environmental in nature, and only a handful of about 90 species are frequent human pathogens. Some of these human pathogens, such as the adult roundworm, pinworm, guinea worm (called the "fiery serpent" in the Bible), and tapeworm, are large enough to be seen with a naked eye. Accordingly, they have been described even in the ancient Indian, Egyptian, and Chinese medical texts, and later by Ancient Greek and Roman physicians. It was only after the discovery of the microscope in the 17th century that other parasites started to be identified, particularly the protozoa and the larval forms of helminths. The 18th and 19th centuries saw the microscopic descriptions of numerous parasites, including their taxonomical classifications. Starting with the early 20th century, we have witnessed the gradual rise of immunological methods being brought into use to diagnose parasitic infections. Perhaps the first such test was the complement fixation test by Pietro Ghedini in 1907 used to diagnose hydatid disease. The accuracy and efficacy of this test, and that of the previously discovered agglutination test by Georges-Ferdinand Widal for typhoid fever in 1896, encouraged scientists to develop and use these immunological methods for other parasites, with varying levels of success. A further revolution started during the closing years of the last century in the form of the amplification of genes and their identification by the novel technique of polymerase chain reaction. This system has been found to be exquisitely sensitive and highly specific for identifying any parasite or other organisms. Its popularity and widespread use in the present century have resulted in pushing the other modalities of diagnosis to the background and threatening to displace microscopy or culture as the gold standards for identification and classification of organisms.

Microscopy remains the preferred tool for diagnosis of parasitic infections throughout the world, while nucleic acid amplification and detection remain a distant dream for the great majority middle- and lower-middle-income countries in Africa, Southeast Asia, and several countries in Central and South America where parasitic diseases are rampant. In such a scenario, adoption of a middle path in the form of immunological diagnosis can be the preferred choice. It has a clear edge over microscopy in terms of sensitivity and over molecular methods in terms of cost effectiveness. Rapid point-of-care antigen or antibody detection tests are particularly suited for remote field areas, where installation and maintenance of high-end instruments are an impractical proposition. This calls for an urgent need to develop newer immunological tests with sensitivity and specificity to rival nucleic acid amplification tests in addition to reevaluating the older, time-tested techniques, which are cheap and easy to perform without the need of sophisticated equipment.

This book has been designed to represent a healthy balance between a theoretical treatise and a procedural handbook. The principle and the important practical considerations of commonly employed immunological tests have been supplemented with detailed protocols for designing in-house or experimental tests, curated from select, peer-reviewed journal articles and online resources from leading manufacturing institutions. Cutting across disciplines, it is envisaged that *Parija's Immunologic Diagnostics in Parasitology: Principles, Techniques, and Applications* will be useful to anyone interested in parasitic disease diagnostics. The book aims to provide the necessary understanding of principles of immunological tests and step-by-step instructions for designing tests in the laboratory. We hope that it will be of particular help to postgraduate students and research scholars, in addition to faculties and scientists across the disciplines of medical and veterinary sciences.

We wholeheartedly thank all the authors of the research articles from which we have derived the test protocols mentioned in this book. Our publisher, CRC Press (imprint of Taylor & Francis Group), has rendered immense help and cooperation in executing this project at every step. We hope this book will also have an international readership across all continents for those interested in knowing, understanding, and developing an immunodiagnostic method for parasitic infections.

Subhash Chandra Parija
Pondicherry, India

Abhijit Chaudhury
Tirupati, India

Feedback Form

Dear Reader,

We request your help in improving this book, *Parija's Immunologic Diagnostics in Parasitology: Principles, Techniques, and Applications*. Your valuable response and feedback will go a long way in achieving our objective of writing this book as an informative guide for understanding the background of various tests and practical considerations while developing such tests at your laboratory.

1. On a scale of 1–10, how useful did you find the book?
2. Are there procedures discussed in this book that have not been presented adequately or are confusing? Please list them and send the modifications necessary to improve these procedures with appropriate references.
3. Are there any other procedures not discussed in the book that need to be added? Please mention them with explanations.
4. Have you found any errors in any part of the book? Please describe them.
5. Any additional comments?

We thank you for your valuable input. Please send your response to us and we will get back to you.

Prof. Subhash Chandra Parija
subhashparija@gmail.com

Dr. Abhijit Chaudhury
ach.patna@gmail.com

Part I

Principles of Immunodiagnostics in Parasitic Infections

1 Introduction to Parasitology

INTRODUCTION

Medical parasitology is the study of animal parasites that infect and produce disease in animals. The world of parasites causing human diseases encompasses single-celled protozoa, multicellular helminths, and the arthropod ectoparasites. Although most of the 65,000 protozoa and more than 75,000 helminth species are free living in nature, some of them can parasitize plants, invertebrates, and vertebrate animals including humans. The human species during its relatively short history on earth have become hosts to about 300 species of helminths and more than 70 protozoan species, and though many of them are rare and accidental, about 90 species are frequently seen to cause human infections. The *Homo sapiens* have acquired them from their primate ancestors and the animals they have domesticated over the years.

THE PARASITE

A *parasite* is an organism that lives on or inside a host and derives biological advantage—such as nutrients, shelter, or reproduction—at the host's expense. Parasites depend on the host for survival and often cause varying degrees of harm, ranging from mild irritation to severe disease. In medical parasitology, parasites include protozoa, helminths, and arthropods, each with distinct life cycles and modes of transmission.

Limitations of Koch's Postulates in Parasitic Diseases

Koch's postulates were designed for bacterial pathogens. Parasites, however, follow different biological rules. Several features make the classical framework unsuitable. These include:

1. **Difficulty in culturing parasites:** Many parasites cannot be maintained in artificial media. For example, *Plasmodium* species need human or mosquito tissues for development and cannot be cultured through all stages *in vitro*.
2. **Complex life cycles:** Parasites often require multiple hosts or vectors. This makes "isolation" and controlled experimentation difficult. For example, the life cycle of *Schistosoma*spp. includes humans and freshwater snails, with separate developmental phases in each host.
3. **Ethical barriers:** Human infection cannot be reproduced for experimental purposes. For example, deliberate infection with *Echinococcus granulosus* larvae is neither feasible nor ethical.
4. **Chronic or intermittent illness:** Parasites can cause prolonged or fluctuating disease. This blurs the causal link between the organism and symptoms.For example, *Trypanosoma cruzi* infection may remain asymptomatic for years before causing cardiac or gastrointestinal disease.

Need for Modified Causation Frameworks

Parasitic diseases involve interaction among parasite, host, vector, and environment. Clinical features may mimic other infectious or noninfectious conditions. Global travel, climate shifts, and ecological changes continue to expand the distribution of many parasites. These trends highlight

DOI: 10.1201/9781003644224-2

the need for reliable, parasite-specific causation criteria. To address this gap, Parija and Khurana proposed a revised framework that aligns with modern diagnostics, molecular tools, and the unique biology of parasites.

Parija–Khurana Criteria for Establishing Causation in Parasitic Diseases

Consistent Detection

The parasite—or its antigen or nucleic acid—should be repeatedly identified in relevant clinical samples using validated methods: for example, detection of *Giardia lamblia* cysts or its DNA in stool samples from patients with persistent diarrhea.

Clinical Correlation

The presence of the parasite should match a recognizable clinical pattern. This pattern must correspond to the parasite's stage and the host's immune response.For example, *Leishmania donovani* correlates with fever, hepatosplenomegaly, and pancytopenia in visceral leishmaniasis.

Experimental Infection

Infection in a suitable animal model should reproduce key human disease features when ethically and practically feasible. For example, experimental infection of mice with *Toxoplasma gondii* reproduces tissue cyst formation and helps study disease mechanisms.

Therapeutic Response

Administration of effective antiparasitic therapy should result in reduction or elimination of the parasite and improvement in clinical symptoms: for example, rapid reduction in *Plasmodium falciparum* parasitemia and resolution of fever after artemisinin-based therapy.

Host–Parasite Interaction

Host genetic factors should be evaluated for their role in susceptibility and disease severity. For example, people with certain HLA types show different outcomes with *Leishmania* infection.

Diagnostic Reliability and Molecular Epidemiology

Diagnostic methods must have high specificity and sensitivity. Molecular tools should clarify parasite strains, transmission dynamics, and emerging resistance. For example, PCR genotyping of *Entamoeba histolytica* helps distinguish it from the nonpathogenic *E. dispar*.

There are several parasites that are truly anthroponotic and found to cause only human infections, in contrast to the zoo anthroponotic parasites that are of animal origin but capable of causing human infections. The human-to-human or animal-to-human transmission may be mediated by various animate and inanimate agents, including biological vectors, other arthropods, water, and food, among others.

Both protozoan and helminthic parasites are capable of asexual and sexual reproductive cycles, and based on the cycle present in a particular host, two principal types are recognized. The definitive host is one where the parasite undergoes the sexual cycle, while the asexual lifecycle is completed in the intermediate hosts. Humans are sometimes the accidental hosts of animal parasites, and in many such cases humans serve as the dead-end host when the parasite is unable to infect other humans or animals.

Geographical Distributions of Parasites

Parasitic infections have a global presence, although many of them are more prevalent in areas with low socioeconomic status and poor sanitation (Figure 1.1). The soil-transmitted helminths comprise a group of nematode parasites and are transmitted by animal and human fecal contamination of soil

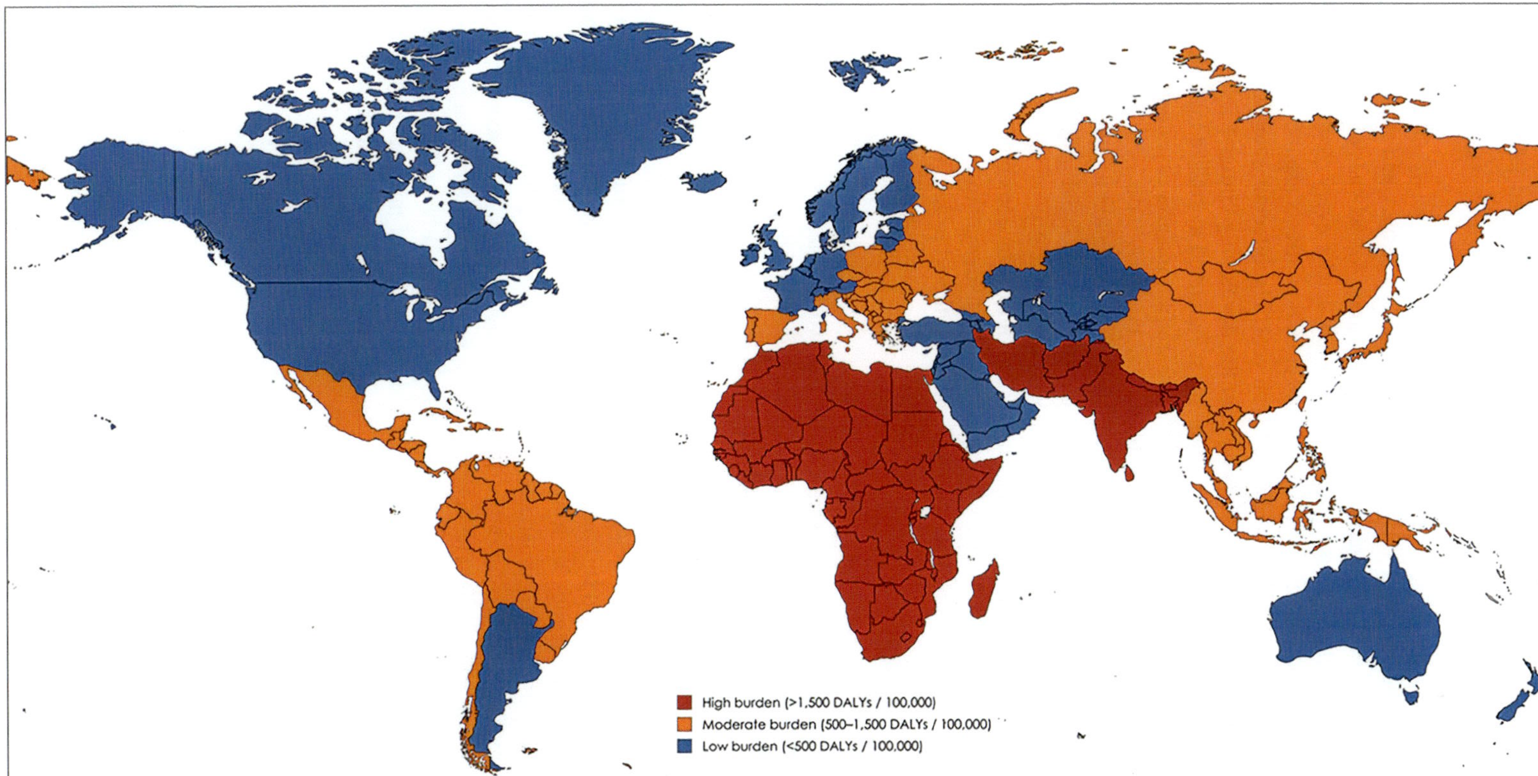

FIGURE 1.1 Global burden of parasitic diseases. This figure illustrates the heterogeneous global distribution of the parasitic disease burden across indicative disability-adjusted life year (DALY) categories. Regions with a high burden (>1,500 DALYs per 100,000 population) are predominantly concentrated in tropical and subtropical settings, reflecting sustained transmission of protozoan and helminth infections under conditions of poverty, limited sanitation, and high vector density. Moderate-burden regions (500–1,500 DALYs per 100,000) represent areas undergoing epidemiological transition, where control efforts have reduced but not eliminated transmission. Low-burden regions (<500 DALYs per 100,000) correspond largely to temperate, high-income settings with effective public health infrastructure. The figure provides a comparative, conceptual overview rather than country-level estimates and is intended to contextualize the persistent global health and equity challenges posed by parasitic diseases.

(The figure is copyright free and has been created using mapchart.net.)

in areas with poor sanitation. These infections affect the people with low socioeconomic status and most deprived communities in Africa, South America, and Southeast Asia. The parasitic infections are one of the most common infections on a global scale, affecting an estimated 1.5 billion people or 24% of the world's population.

Burden of Diseases

Globally, almost 3.5 billion people are infected by intestinal protozoan parasites, and approximately 1.7 billion annual diarrheal disorders occur due to protozoal infections. Amoebiasis affects 10% of the global population and is thought to cause 40,000 to 100,000 deaths every year worldwide. According to the World Health Organization (WHO), *Giardia* infection affects around 200 million people worldwide annually, with 500,000 deaths reported each year. The infection rates for *Giardia* stands at 2%–5% in the developed nations and 20%–30% in the developing countries. Two other protozoan diseases merit mention. The first is malaria, which has remained a scourge in Africa and Southeast Asia. Globally in 2022, there were an estimated 249 million malaria cases and 608,000 malaria deaths in 85 countries. The African region alone was home to 94% of malaria cases (233 million) and 95% (580,000) of malaria deaths of which 80% were children under 5 years of age. The other is leishmaniasis in its various forms in Asia, Africa, and South America. An estimated 700,000 to 1 million new cases of leishmaniasis occur annually, of which 50,000 to 90,000 are of visceral leishmaniasis, and the rest of cutaneous and mucocutaneous types (WHO Fact sheets, 2023). Although African sleeping sickness has been brought under control, Chagas disease still largely remains endemic in South America, with 6–7 million infected people to date.

LABORATORY DIAGNOSIS

With the backdrop of the tremendous burden of parasitic diseases worldwide, it is natural to assume that there is a repertoire of diagnostic tests available for these infections; but ironically, there is a huge gap in the development and availability of diagnostic tests for parasitic infections.

Microscopy

Ever since Leeuwenhoek observed the first protozoan parasite *Giardia* in his own stool specimen in 1674 with his crude microscope, the world still relies on the light microscope for parasitic diagnosis. Notwithstanding its inherent shortcomings such as low sensitivity, operator variability, need for multiple specimen (particularly for intestinal parasites), sometimes invasive techniques needed for specimen collection, and misdiagnosis arising out of similar looking artifacts, microscopy remains the gold standard in most parasite diagnosis. The development of new techniques for diagnosis—be it immunologic, genomic, or proteomic based—still needs to be compared favorably with microscopy to have credibility in the scientific community. The main advantage of microscopy lies in its simplicity. A good light microscope, suitable specimen, a few staining reagents, and an experienced microscopist are all that is needed to give a reliable identification of the parasite.

Immunodiagnosis

Immunodiagnosis of parasitic infections uses host antibody or antigen detection to support rapid and specific identification when parasites are scarce or difficult to isolate. Tests such as ELISA, rapid immunochromatographic assays, and molecular-linked immunoassays help confirm exposure, active infection, or disease severity. These tools improve sensitivity, guide early treatment, and aid in monitoring therapy response. Their accuracy depends on the stage of infection, the host's immune status, and the quality of test antigens, so the results are best interpreted alongside clinical and epidemiological data.

The immunodiagnostic encompasses intradermal skin tests and serological tests.

Intradermal Skin Tests

Intradermal skin tests assess delayed-type hypersensitivity to parasite antigens by injecting a small amount of standardized antigen into the skin and observing the local reaction. A positive wheal and erythema, usually appearing within 24–48 hours, suggests prior sensitization rather than active infection. These tests can support the diagnosis of a few parasitic infections such as hydatid disease or toxoplasmosis but have limited specificity and may remain positive long after exposure. They are adjunct tools, and as such, they need to be interpreted with clinical findings and other laboratory tests.

Serological Tests

Serological tests began with the development and evaluation of the Widal agglutination test for typhoid fever in 1896. This was subsequently followed by the development of a complement fixation test (CFT) for antibody detection in 1901 by Bordet and Gengou. This technique was soon put into practice by August von Wassermann in 1906 when he developed the first serological test for syphilis. The 20th century has seen intense efforts by research scholars and scientists to develop antigen- or antibody-based serological tests for various parasitic diseases, both human and animal. All available formats have been experimented with, and although many of them might have been promising, only a handful has seen the light of day in the form of widely available commercial test kits.

Many advanced research laboratories and institutions in India and abroad have been making a consistent effort to design and evaluate in-house serological tests but not available for commercial use. A few reliable antibody-based and antigen-based serological tests are available for important parasitic diseases such as malaria, toxoplasmosis, amoebiasis, giardiasis, cryptosporidiosis, filariasis, and toxocariasis, among others. Serological tests have the advantage of rapid turnaround time, being objective without inter-operator variability, and ease of performance, but in many cases cross-reactivity with related species or genera poses a problem. The increasing use of proteomic studies of different parasites may pave the way to discovery of unique molecules that can be used to develop highly specific immunological assays for important parasites.

GENOMICS AND PROTEOMICS

The 21st century has seen tremendous advancements in genomic and proteomic studies of a wide variety of pathogens, including parasites. The discovery of species- or genus-specific genes has paved the way for its amplification by various techniques, resulting in accurate identification of a parasite. Interspecies discrimination of the parasites has been made easy by gene sequencing and in more recent years by whole-genome sequencing, which is steadily becoming more affordable.

Various tests based on nucleic acid amplification have much higher sensitivity compared with microscopic or serologic tests. These tests can detect very small numbers of parasites in a sample. They can also detect multiple genetic sequences simultaneously, leading to the identification of several parasites at one go. Despite the many advantages of nucleic acid–based tests in clinical parasitology, however, in many instances the gene targets used may not be highly specific for a given species or genus. One such instance that has been recently described concerns the controversial protozoan parasite *Lophomonas blattarum* associated with respiratory tract infections where it has been pointed out that the primers used for its identification can also amplify genes from other unrelated species. Nevertheless, nucleic acid amplification assays appear poised to endure and may, over time, establish themselves as the definitive reference method for parasite identification, displacing traditional microscopy that has dominated for centuries.

The advent of proteomic studies has opened a new horizon in microbial diagnostics. Mass spectrometry had been a part of chemical sciences since the 19th century, but it was only in the

1980s and 1990s that this technique was applied to detect oligonucleotides and proteins of microorganisms. The introduction of Matrix Assisted Laser Desorption Ionization – Time of Flight (MALDI-TOF) technology is gradually replacing the time-consuming conventional tests and nucleic acid techniques in bacterial and fungal identification. The availability of automated instrumentation has resulted in rapid identification within 45–90 minutes in culture-grown organisms. The development has resulted from the generation of a large database of protein profiles of numerous bacteria and fungi. On the other hand, in virology and parasitology, MALDI-TOF remains a promising technology that needs further exploration. A derivation of MALDI-TOF is the Surface Enhanced Laser Desorption Ionization–Time of Flight (SELDI-TOF) platform, which has been used for parasites. Since culture techniques for parasites are mostly nonstandardized and time consuming, direct detection from clinical samples is desirable. However, this requires purification of parasite stages such as cysts, trophozoites, or helminths from complex specimen like stool and sputum, which is a lengthy procedure, in addition to the limited availability of proteomic databases of commonly encountered parasites.

CONCLUSION

An analysis of the current literature on parasitic diagnostics reveals that microscopy remains the first choice but is being gradually supplemented with immunological tests and supported further by nucleic acid amplification assays and sequencing of specific genes for finer distinction of genus/species. Occurrence of antibodies is common in endemic parasitic diseases even in the absence of any recent or ongoing infection, and therefore antigen detection has a greater role to play in these cases. Currently, the parasite proteomics is still in its infancy, but it holds great promise in the development of improved diagnostics in the near future.

FURTHER READINGS

Cox FE. History of human parasitology. *Clin Microbiol Rev.* 2002;15:595–612.

Disease information sheet. Available at: https://www.pasteur.fr/en/medical-center/disease-sheets/Amebiasis. Accessed April 17, 2026.

Mewara A, Gile GH, Mathison B, Zhao H, Pritt B, Bradbury RS. *Lophomonas* as a respiratory pathogen: jumping the gun. *J Clin Microbiol.* 2024;62:e0084523.

Parija SC, Khurana S. Revisiting Koch's postulates: a tailored approach for clinical parasitology. *Trop Parasitol.* 2025;15:8–11.

WHO Fact sheets. Available at: https://www.who.int/news-room/fact-sheets/detail/malaria. Accessed April 17, 2026a.

WHO Fact sheets. Available at: https://www.who.int/news-room/fact-sheets/detail/leishmaniasis. 2023. Accessed April 17, 2026.

WHO Fact sheets. Available at: https://www.who.int/news-room/facts-in-pictures/detail/chagas-disease. Accessed April 17, 2026b.

World Health Organization. Soil-transmitted helminth infections: key facts: World Health Organization. Available at: https://www.who.int/news-room/fact-sheets/detail/soil-transmitted-helminth-infections. Accessed October 7, 2025.

Yusuf E, Allerberger F. Infectious disease serology in 2021. *Clin Microbiol Infect.* 2021;27:1204–1206. doi:10.1016/j.cmi.2021.06.020

2 Immunodiagnostics of Parasitic Infections

INTRODUCTION

The foundation for the modern science of immunology was laid by Emil von Behring and fellow researcher Kitasato in 1890 when they discovered that infected animal serum contains substances that have antibacterial activity, particularly bacterial toxins. Behring successfully translated his research in the development of diphtheria antiserum to be used for serum therapy. During the same period, in 1897, Paul Ehrlich proposed the side chain (later the term was changed to receptors) theory arguing that for the toxins (antigens) generated during infection, specific receptors (antibodies) are produced by the cells that are secreted into the serum and neutralize the toxins. This laid the foundation for the concept of antigen antibody interaction. In England, Almroth Wright demonstrated the action of antibodies in bacterial phagocytosis. Hans Buchner in Germany in 1889 and Jules Bordet in Belgium in 1896 first described a heat-labile substance from immunized animals and humans and named it Alexine, which was later renamed Complement by Paul Ehrlich. These discoveries paved the way for utilizing the knowledge gained in the development of diagnostic tests for human and animal use.

TIMELINE OF DEVELOPMENT OF IMMUNODIAGNOSTICS

The scientific study of immunology began in the 1890s in Europe with the pioneering work of scientists such as Behring, Ehrlich, Kitasato, Bordet, and others. Simultaneously, the new immunological principles that were being developed found their way into the development of diagnostic tests for infectious and noninfectious diseases. Figure 2.1 shows the timeline for the introduction of immunological tests for parasitic diagnosis.

Immunodiagnostics in parasitic diseases encompasses intradermal skin tests and antigens- and antibodies-based serological tests.

Intradermal Skin Test

Intradermal skin test is an approach that is totally different from the existing immunodiagnostic techniques being discussed later, but it had been used widely in the past. One such early test was Casoni's test for hydatid disease first developed in 1912. Similar tests came into use in subsequent years for leishmaniasis, schistosomiasis, trichinellosis, paragonimiasis, and some other parasitic diseases. However, with the advent of newer, safer, and more sensitive and specific *in vitro* tests, the intradermal test gradually fell into disuse.

Serological Tests

Complement Fixation Tests

Complement fixation test (CFT) was one of the earliest tests introduced in diagnostic parasitology. The discovery of complement and its properties prompted Bordet and Gengou in 1901 to use it for typhoid diagnosis, followed by Wassermann in 1906 for syphilis.

DOI: 10.1201/9781003644224-3

Timeline of Immunodiagnostic Techniques

The following chart presents the chronological development of major immunodiagnostic techniques.

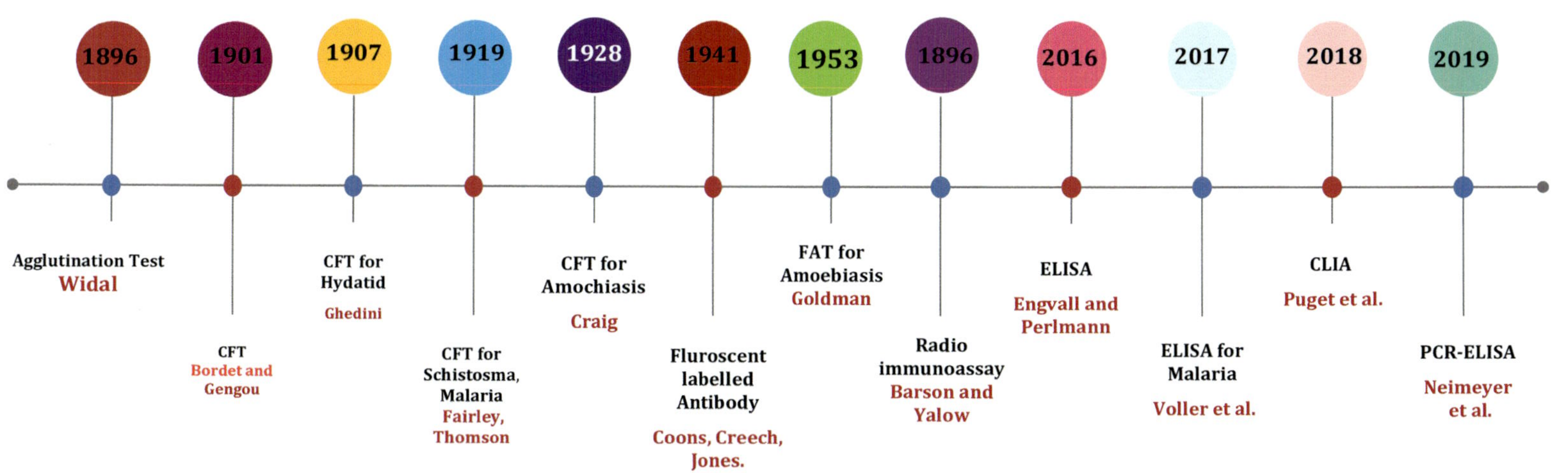

FIGURE 2.1 The timeline of development of major immunodiagnostic techniques.

Ghedini in 1907 first used it for serological diagnosis of hepatic hydatid and investigated *Ancylostoma* and *Ascaris* infestations; this was later refined by Weinberg and Fairley. In 1913, a technique for growing malaria parasites was discovered, which was used for antigen preparation and subsequently employed for malaria diagnosis by CFT in 1919. In the same year, Fairley described a complement fixation test for Bilharziasis using antigen from infected snail liver containing cercariae of *Schistosoma mansoni*. This work was based on the study by Tanaka from Japan in 1912 who had used cercarial extract of *Schistosoma japonicum* for detecting antibodies in rats using the complement fixation method. Izar in 1914 first described complement fixation for *Entamoeba histolytica*, and in 1928 Charles F. Craig first developed a test based on this principle, which could be adapted for routine diagnosis of amebiasis. The subsequent decades until the 1970s saw the increasing use of this test for various other parasitic infections such as leishmaniasis, malaria, trypanosomiasis, schistosomiasis, filaria, and others, and for a while it was considered the standard reference test for several infections. However, with the advent of newer tests, it gradually lost its importance because it was labor intensive, cumbersome, and difficult to standardize.

Agglutination and Precipitation Tests

The history of immunodiagnostics began with the first discovery of agglutination test for diagnosis of typhoid fever by Georges-Ferdinand Widal in 1896. Agglutination tests and precipitation techniques attracted the attention of early investigators, since they were easy to perform and standardize. Unlike the CFT, some forms of agglutination tests are still in active use, as evidenced by the Widal test, which is still used in many countries 130 years after its discovery. Herbert Edward Durham and Max von Gruber first described the agglutination phenomenon in 1896, which was then termed *Gruber-Durham reaction*. Gruber proposed the term *agglutinin* (antibody) for the substance that causes agglutination of cells.

The use of red blood cells as carriers of antigen in the format of indirect hemagglutination test (IHA) was widely used for trichinosis, echinococcosis, amoebiasis, toxocariasis, and toxoplasmosis in the second half of 20th century. In addition to the popular IHA technique, other formats like bentonite flocculation, latex agglutination, and carbon particle agglutination were evaluated for several helminthic and protozoan parasites, with variable results. In 1897, Rudolf Kraus introduced the fluid precipitin test using culture filtrate of the cholera bacilli. This technique was then adopted in several formats and used for various parasitic diseases. However, among these tests, only the circumoval precipitin test—first reported by Oliver-Gonzales in 1954 for the diagnosis of schistosomiasis—stands out as a truly time-tested method.

Fluorescent Antibody Technique

In 1941, Coons, Creech, and Jones from Harvard University reported the use of fluorescent labeled antibodies to demonstrate pneumococcal antigens in infected tissue specimens. This discovery opened a new vista in immunodiagnostics, paving the way for development of newer techniques in immune assays using radionuclide and enzyme labels. Although it did not receive much attention in the 1940s, in the following decade the fluorescent antibody technique (FAT) was applied in detection of antigens as well as antibodies by a wide range of researchers including parasitologists. Goldman can be credited with the application of immunofluorescent test for the first time for parasitic diagnosis in 1953 when he described the differentiation of two species of *Entamoeba*, using specific fluorescent antibodies. Later he extended his work to develop fluorescence inhibition test for detection of *Toxoplasma* antibodies. The first report of its use in helminthology was a report by G. J. Jackson in 1959 on his work with *Trichinella spiralis*, which was successfully translated for detection of schistosomal antibodies in patient serum using cercarial antigen by Sadun and coworkers. At present, several commercial FAT diagnostic kits are available for both antigen and antibody detection for several parasites, which will be discussed in the subsequent chapters.

Radioimmunoassay

The work of Barson and Yalow dealing with the metabolism of ^{131}I-labeled insulin in diabetes was the starting point in the development of radioimmunoassay (RIA) as an attractive technique for detection of antigens and antibodies. During the initial period, the RIA was used mainly to estimate the extremely low amounts of peptide hormones in blood present in concentrations of 10^{-10}–10^{-12} moles. It was first used in infectious diseases in 1970 when Walsh, Barson, and Yalow used it to detect Australia antigen (Hepatitis B surface antigen) and antibody. In the two decades between 1970 and 1990, there was some interest in this technique in parasitology for malaria diagnosis, but it was eclipsed by the development of enzyme immunoassays during the same period.

Enzyme-Linked Immunosorbent Assay

The introduction of enzyme-linked immunosorbent assay (ELISA) in 1971 revolutionized the field of immunoassays in infectious disease diagnostics. In that year Peter Perlmann and Eva Engvall in Sweden reported the quantification of IgG in rabbit serum using alkaline phosphatase enzyme as a reporter label. In the same year, Anton Schuurs and Bauke Van Weemen in the Netherlands demonstrated an immunoassay using horseradish peroxidase enzyme coupled with glutaraldehyde. It is interesting to note that parasitologists were the first to use this technique for diagnosis of malaria and trichinellosis, and it was soon extended to toxoplasmosis and trypanosomiasis. At present, it is one of the most common methods for antigen and antibody detection for various parasitic diseases.

A non-ELISA immunoassay using a luminol-fusion target protein was developed in 1977 by Puget and colleagues in France. This innovation served as the forerunner of the chemiluminescence immunoassay (CLIA) technique, which matured in the 1980s. At present, several CLIA kits are commercially available, beginning with *Toxoplasma* IgM kit in 2002. It includes CLIA for neurocysticercosis, Chagas disease, liver fluke infections, giardiasis, and other parasitic infections.

Point-of-Care Tests

The one-step pregnancy test developed in the 1980s was a revolutionary concept for rapid, easy-to-use, accurate point-of-care testing for antigen or antibody, and it was quickly adopted for infectious disease testing. Beginning with the introduction of malaria detection kits in the early 1990s, many such kits are commercially available at present for amoebiasis, giardiasis, cryptosporidiosis, filariasis, Chagas disease, and others. Although many of them are not yet U.S. Food and Drug Administration (FDA) approved, they are being widely used in countries where these diseases are endemic.

Nucleic Acid–based Diagnostics

The 21st century has seen a phenomenal growth of nucleic acid–based diagnostics, and in many instances they are becoming the standard tests for microbial diseases, replacing the age-old culture or microscopy tests that have been considered the gold standards. The advent of these tests with exquisite sensitivity and near 100% specificity has overshadowed the immunological tests in recent times, as reflected by the slowdown in the development of new immunological tests in the last two to three decades. Despite the advantages of the nucleic acid–based tests, these are still not widely available in the geographical areas where the parasitic diseases are widely prevalent, including some of the most economically challenged countries of the world in Asia, Africa, South America, and other areas.

CONCLUSION

Many of the immunological tests that had been developed in the past but abandoned over the years, such as agglutination tests, are easy to perform and highly economical. Their lack of sensitivity and/or specificity can be explained by the nonavailability of standardized antigens, cross-reactivity with other related parasites, use of crude antigens, and, in many cases, use of nonspecific antigens. In this modern era with improved access to purified or recombinant antigens and monoclonal antibodies, there is a need to give some of these old tests using specific antigens or antibodies another look. This would pave the way for the development of economical, easy to use in field conditions, rapid tests in countries with limited access to sophisticated laboratories and inadequate trained manpower.

FURTHER READINGS

Berson SA, Yalow RS, Bauman A, Rothschild MA, Newerly K. Insulin-I131 metabolism in human subjects: demonstration of insulin binding globulin in the circulation of insulin-treated subjects. *J Clin Invest.* 1956;35:170–190.

Casoni T. La diagnosi biologica dell'echinococcosi umana mediante l'introdermoreazione. *Folia Clinica Chimica et Microscopica.* 1912;4:5–16.

Coons AH, Creech HJ, Jones RN. Immunological properties of an antibody containing a fluorescent group. *Exp Biol Med.* 1941;47:200–202.

Craig CF. Complement fixation in the diagnosis of infections with *Endamoeba histolytica. Am J Trop Med.* 1928;8:29–37.

Ehrlich P. Die Seitenkettentheorie und ihre Gegner. *M€unchner Med Wochenschr.* 1901:2123–2124.

Engvall E, Ljungström I. Detection of human antibodies to *Trichinella spiralis* by enzyme-linked immunosorbent assay (ELISA). *Acta Pathol Microbiol Scand Sect C.* 1975;83:231.

Engvall E, Perlmann P. Enzyme-linked immunosorbent assay (ELISA): quantitative assay of immunoglobulin G. *Immunochemistry.* 1971;8:871–874.

Fairley NH. The discovery of a specific complement fixation test for bilharziasis and its practical application to clinical medicine. *J Royal Army Med Corps.* 1919;32:449.

Fairley NH. The complement fixation test for hydatid disease and its clinical value. *Med J Australia.* 1922;9:341–345.

Goldman M. Cytochemical differentiation of *Endamoeba histolytica* and *Endamoeba coli* by means of fluorescent antibody. *Am J Hyg.* 1953;58:319–328.

Puget K, Michelson AM, Avrameas S. Light emission techniques for the microestimation of femtogram levels of peroxidase: application to peroxidase (and other enzymes)-coupled antibody–cell antigen interactions. *Anal Biochem.* 1977;79:447–456.

Sadun EH, Anderson RI, Williams JS. The nature of fluorescent antibody reactions in infections and artificial immunizations with *Schistosoma mansoni. Bull World Health Organ.* 1962;27:151–159.

Sinha S, Kaur U, Sehgal R. Diagnosis of parasitic zoonoses. In Parija SC, Chaudhury A, editors. *Textbook of Parasitic Zoonoses.* Singapore: Springer Nature; 2022:59–74.

Tanaka M. The complement fixation test for *S.* japonicum. *Chiugai Iji Shimpo, Tokio.* 1912;33(2):721.

Thomson JG. Experiments on the complement fixation in malaria with antigens prepared from cultures of malarial parasites (*Plasmodium falciparum* and *Plasmodium vivax*). *Proc R Soc Med.* 1919;12(Med Sect):39–48.

Van Weemen BK, Schuurs AH. Immunoassay using antigen–enzyme conjugates. *FEBS Lett.* 1971;15:232–236.

Voller A, Bidwell D, Huldt G, Engvall E. A microplate method of enzyme-linked immunosorbent assay and its application to malaria. *Bull World Health Organ.* 1974;51:209.

Walsh JH, Berson SA, Yalow RS. Detection of Australia antigen and antibody by means of radioimmunoassay techniques. *J Infect Dis.* 1970;121:550–554.

Wright AE. *Studies on Immunisation and Their Application to the Diagnosis and Treatment of Bacterial Infections.* London: Constable; 1909.

3 Immunological Responses in Parasitic Infections

INTRODUCTION

Any organism entering a host encounters the inherent defense system that tries to ward off the invader. There may be two important outcomes of the interaction between the body's immune system and the organism. The parasite can establish its residence in the host by defying the defense system and initiate infection, or else it may be eliminated without causing any effect on the host. This complex interplay between the host and the pathogen is orchestrated by the innate and adaptive immune systems of the body. The innate system serves as the first line of defense that by itself may be successful in preventing an infection from setting in. In the event of its apparent failure to prevent the entry of the microbe, the adaptive immune system comes into play, and in most instances the dual action is sufficient to control the infection and eliminate the pathogenic microorganism.

INNATE IMMUNE SYSTEM

Physical and biochemical barriers on body surfaces and mucosal linings of the internal body systems like gastrointestinal or respiratory tracts act as barriers for entry of the microorganisms into the body. The skin not only acts as a mechanical barrier but has a direct inhibitory effect due to fatty acids and lactic acid in the secretions. The ciliated epithelium of respiratory tract acts as an effective trapping mechanism for the microbes because of the thick mucus. These trapped organisms are effectively removed by the ciliary action of the epithelial cells, via the mechanical action of coughing or sneezing. Similarly, the extremely high acidity of the stomach can kill many of the microbes ingested with food and water. The commensal organisms present on the skin, in the throat, and in the lower gastrointestinal tract resist the lodging of pathogens by competing with them as well as by the release of acidic molecules and colicins.

In the event the mechanical barriers are unable to prevent the entry of the pathogen into the body, it encounters other defenses. One is the soluble antimicrobial factors (peptides) present in blood and body secretions produced by neutrophils, macrophages, and epithelial cells. It includes the α- and β-defensins as well as human cathelicidin LL37 or human cationic antimicrobial peptide. The enhanced mucus production in the intestine in response to helminth infections is regulated by both innate and adaptive immune systems and helps with rapid propulsion and expulsion of the worms from the system. The other formidable innate response is in the form of cellular attack. The cellular effect is initiated by the pathogen-associated molecular patterns (PAMP) present on the pathogens, which are recognized by the pattern recognition receptors (PRR) on the cells. Various PAMPs such as phosphoglycans, ES-62, mannose binding lectins, and glycophosphatidyl inositol have been identified on *Leishmania*, *Plasmodium falciparum*, and other parasites. The Toll-like receptors (TLR) are the best-known PRRs, apart from NOD-like receptors, cysolic DNA sensors, and others. The PAMPs upregulate the expressions of TLRs resulting in synthesis of proinflammatory cytokines, nitrogen oxide synthase, and other effectors. The important host cells that actively participate in the innate response include macrophages, granulocytes, natural killer (NK) cells, NK-T cells, innate lymphoid cells (ILC), and γ-δ T cells.

The important cells include the following.

DOI: 10.1201/9781003644224-4

Macrophages and Granulocytes

Innate immune response to protozoan parasites depends largely on the phagocytic mechanisms of macrophages and neutrophils. Once phagocytosed by these cells, the ingested parasite may be killed by free radical injury. On the other hand, parasites have their own defense mechanism to evade the intracellular killing. It includes inhibition of respiratory burst mechanisms and escape from phagolysosome once inside the cells. This type of classical macrophage activation and response is shown by the M1 type of macrophages. A second set of macrophages has also been described and named alternatively activated macrophages (AAM or M2). The M2 macrophages are associated with TH2 adaptive immune response seen in helminthic infections. Eosinophils have a role to play in attacking the helminths and damaging their outer membranes by releasing enzymes and peptide molecules present in the granules.

Natural Killer Cells

Along with macrophages and granulocytes, natural killer (NK) cells are particularly active against protozoan parasites, as well as hookworm. NK-cell activation is initiated by PRR interacting with dendritic cells and is a contact- as well as cytokine-driven event. The result is interferon-γ (IFN-γ) production, which is an effector molecule for both innate and adaptive responses. These cells can be considered the innate counterparts of cytotoxic CD8+ T cells of adaptive immunity. Certain cytokines, such as interleukin-10 (IL10) downregulate NK-cell expression which is useful in protecting the host from excessive damage to tissues by IFN-γ.

Innate Lymphoid Cells

Innate lymphoid cells (ILC) is a family of newly described cells that function as T-lymphocytes of adaptive response but lack antigen receptors and are unable to undergo clonal selection. They are capable of various cytokine production after interaction with antigens. Thus ILC1, ILC2, and ILC3 can be considered the innate counterparts of TH1, TH2, and TH17 cells, respectively. ILC1 is active in intracellular protozoan infections while ILC2 is active against helminths.

ADAPTIVE IMMUNE SYSTEM

The T- and B-lymphocytes with initial priming innate immune response are the principal mediators of adaptive response. While the B cells are responsible for antibody production, which has a limited role in protection against parasite infections, the T cells have multiple functions. Toward this end, there are four important subsets of T cells that have effector functions: T-helper cells (TH, CD4+ T cells), cytotoxic T cells (CTL, CD8+ T cells, Tc), T17 cells, and T-regulatory cells (TREG). The T-helper cells are of two types, TH1 and TH2, the former one being active against intracellular protozoan parasites while the latter is differentiated in response to extracellular helminth infections.

The roles played by these cells are described in the following sections.

TH1 Response

The TH1 response has an important role in protection against the pre-erythrocytic stage of *Plasmodium*, as well as for *Leishmania* and *Toxoplasma* infections. The IFN-γ produced by activated TH1 cell binds to macrophages that produce antiparasitic molecules. It enhances the expression of MHC 1 to help the cytotoxic T cells in their parasiticidal activity, and it increases the expression of MHC II on macrophages to promote antigen presentation to CD4+ T cells.

TH2 Response

The antibody response is elicited in all blood- and tissue-dwelling parasites, but only in a few cases do these have any protective activity. Some degree of protection has been demonstrated in *Toxoplasma*, *Trypanosoma brucei*, and *Plasmodium falciparum*, but most often these weak responses are not enough to eliminate the parasites (Figure 3.1). On the other hand, this response is predominant in helminth infections, and activation of TH2 cells leads to the expression of several cytokines. These include, among others, IL3, IL4, IL5, IL9, IL10, and IL12. IL4 is particularly effective in immunoglobulin class switching to IgE. IgE plays a role in intestinal helminth infections in mediating intestinal anaphylaxis due to mast cell degranulation. This results in rapid elimination of the larval stage of the worms. IL3 and IL9 cause accumulation of mast cells in intestine in response to helminth infections. The mast cells prevent adhesion and penetration of the helminth into intestinal mucosa by producing chondroitin sulphate, in addition to expressing IgE receptors. Similarly, eosinophils are attracted to the site of helminthic infection by IL4 and IL13, and degranulation of eosinophilic granules can lead to parasite destruction, as has been demonstrated by *in vitro* experiments.

T17 Cell Response

TGF-β and IL16 can cause differentiation of T cells into T17 cells, which can produce IL17, a proinflammatory cytokine. This cytokine can help in recruitment of granulocytes at the site of helminth infection and release of other proinflammatory cytokines. The ensuing inflammatory response is detrimental to the parasite and concomitantly causes host tissue damage and resultant pathology. It is to be noted that IL17 is also produced by NK cells and γδ-T cells as part of the innate immune response.

T-regulatory (TREG) Cells

TREG cells are part of the T-lymphocyte family and have the special property to suppress the function of other lymphocytes, including CD4+ and CD8+ T cells as well as B cells. These cells do not have any protective action; rather, they contribute to the chronicity of infection shown by many helminths. Due to their suppression activity, they induce a state of profound immune tolerance in the host. Therefore, the parasite establishes an environment with low parasite-specific lymphocyte proliferation, higher IgG4 concentration compared to IgE, and increased levels of IL10 and TGF-β, with the resultant minimal damage to the host. The parasite can then survive in this safe niche for months or even years.

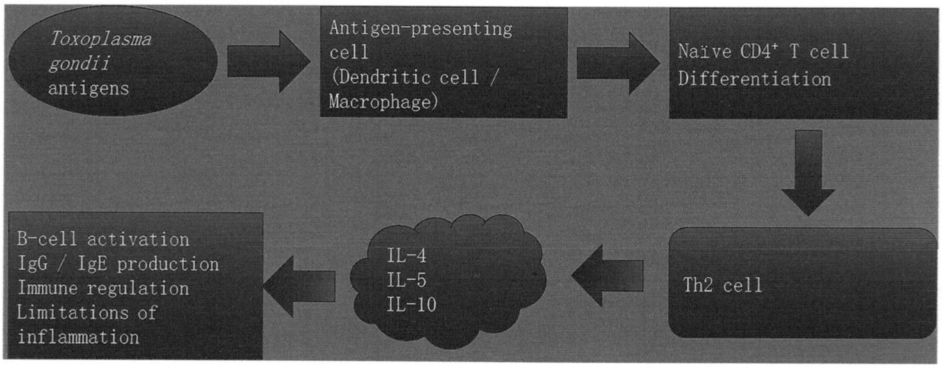

FIGURE 3.1 Th2 immune response in toxoplasmosis. Following antigen presentation, $CD4^+$ T cells differentiate toward a Th2 phenotype, producing IL-4, IL-5, and IL-10 that promote humoral immunity and immune regulation. This response contributes to host–parasite equilibrium, particularly during chronic infection.

IMMUNE EVASION PHENOMENON

This is the phenomenon shown by multiple protozoan and helminth parasites whereby they can evade the effective immune response of the host, with survival benefit. The following sections describe the important evasion strategies used by parasites.

Antigenic Variation

This is the change in the immunodominant antigen of a pathogen due to alteration of the peptide or polysaccharide making up the antigenic molecule. The change may be loss or gain of the antigen or some alteration in the antigenic structure. Therefore, the host immune system is not able to identify the antigen and is ineffective, although it may have been effective against the previous antigenic type. This type of variant-specific surface groups (VSG) has been described in *T. brucei* (Figure 3.2), *Giardia*, and malaria parasites encoded by *VSG genes*. Numerous such genes may be present in a parasite, but only one is expressed at a time, particularly when the antibody level against the previously expressed antigen type reaches a critical level.

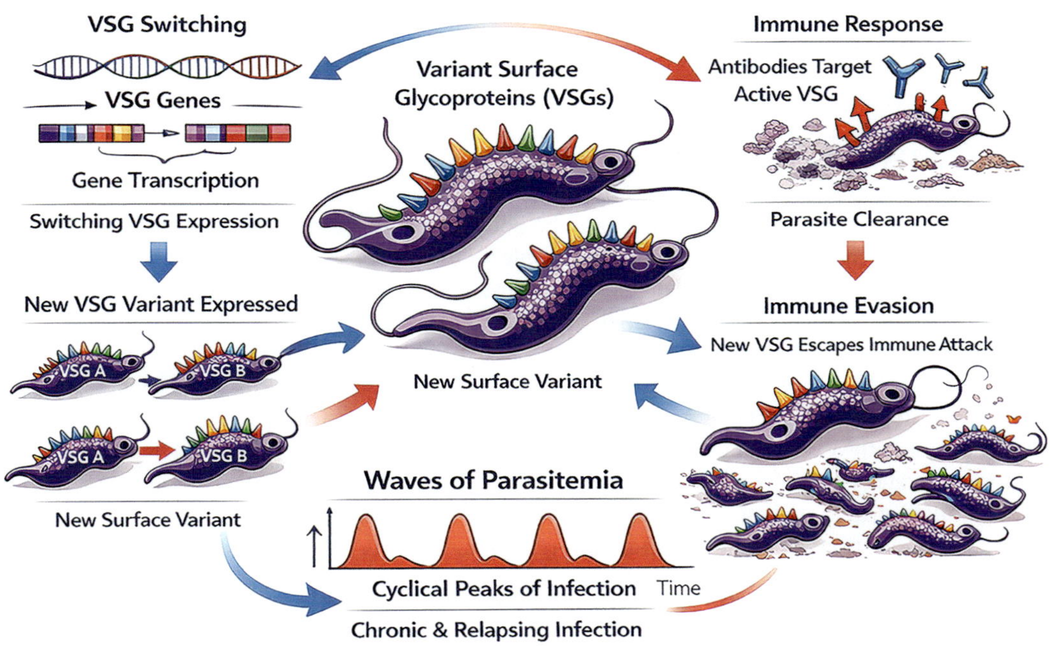

FIGURE 3.2 Antigenic variation in African trypanosomes. A schematic representation of the molecular and immunological bases of antigenic variation in *Trypanosoma* spp. Parasites are coated with a dense monolayer of variant surface glycoproteins (VSGs), of which only a single VSG gene is actively transcribed at any given time from a specialized expression site. Antigen switching occurs through transcriptional switching or recombination-mediated replacement of the expressed VSG gene, resulting in the surface display of a new antigenic variant. Host antibodies efficiently recognize and clear parasites expressing the previously dominant VSG, leading to a transient reduction in parasitemia. Rare parasites that have switched to an alternative VSG escape immune recognition and undergo clonal expansion, reestablishing infection. Repeated cycles of immune clearance and VSG switching generate characteristic waves of parasitemia, underpinning chronic and relapsing infection in the mammalian host.

(The figure is copyright free and generated using an AI-assisted graphic illustrator.)

Immunosuppression

This feature is mainly seen in helminth infections where immune tolerance of the host for the parasite is manifested by parasite survival and chronicity of infection with little damage to the host. Helminth products can induce TGF-β receptor production with the activation of TREG cells as mentioned earlier. TREG cells can suppress the macrophage activity and differentiation of T cells, leading to an immunosuppressive effect. Parasite products also modulate TH-cell differentiation and class switching in B cells.

Molecular Mimicry

The display of certain antigens by the parasite, which have a resemblance to some host molecule, confers a survival advantage for the parasite, as they may not be attacked by the host immune system, considering them to be self-antigens. Thus, *Trypanosoma cruzi* may express some antigen with similarity to fibronectin cell receptors.

Apart from these well-defined mechanisms, such factors as large size of a parasite (helminths), coating with the host cell protein (*Schistosoma*), anatomical seclusion (malaria parasite in RBC), or production of enzymes that can resist immune-mediated damage are some of the other strategies adopted by parasites for immune evasion.

CONCLUSION

The host–parasite interaction for protozoan and helminth parasites is a highly complex phenomenon due to the vast repertoire of antigenic molecules produced by these organisms, as well as the intricate and extensive crosstalk between the innate and adaptive immune responses. Physical and biochemical barriers on body surfaces and mucosal linings of the internal body systems act as barriers for entry of the microorganisms into the body. Once the mechanical barriers fail to prevent the entry of the pathogen into the body, it encounters other defenses such as antimicrobial peptides and host cells that participate in an innate response. These cells include macrophages, granulocytes, NK cells, NK-T cells, ILC, and γ-δ T cells. In the adaptive response, the TH1 and TH2 cells are important players, the former being active against intracellular protozoan parasites while the latter responds to extracellular helminth infections. In addition, T17 cells can produce a proinflammatory cytokine (IL17) that represents a double-edged sword, being detrimental to the parasite but concomitantly causing host tissue damage and resultant pathology. Immune evasion is an important strategy used by many parasites and includes antigenic variation, immunosuppression by T-regulatory cells, and molecular mimicry. Despite tremendous advances in immunology, many unanswered questions remain in parasite immunology, and it is expected that future research will elucidate them.

FURTHER READINGS

Chaudhury A. Parasite immunology. In Parija SC, Chaudhury A, editors. *Textbook of Parsitic Zoonoses*. Singapore: Springer Nature; 2022:21–37.

Ezema CA, Okagu IU, Ezeorba TPC. Escaping the enemy's bullets: an update on how malaria parasites evade host immune response. *Parasitol Res*. 2023;122:1715–1731.

Harris NL, P'ng L. Recent advances in type-2 cell-mediated immunity: insights from helminth infection. *Immunity*. 2017;47(6):1024–1036.

Maizels RM, McSorley HJ. Regulation of the host immune system by helminth parasites. *J Allergy Clin Immunol*. 2016;138:666–675.

McGuinness DH, Dehal PK, Pleass RJ. Pattern recognition molecules and innate immunity to parasites. *Trends Parasitol*. 2003;19:312–319.

Morales-Montor J, McKay DM, Terrazas LI. Advances in the immunobiology of parasitic diseases. *Pathogens*. 2022;11:811.

Tormo N, del Remedio Guna M, Fraile MT, Ocete MD, Garcia A, Navalpotro D, et al. Immunity to parasites. *Curr Immunol Rev*. 2011;7:25–43.

Yasuda K, Nakanishi K. Host responses to intestinal nematodes. *Int Immunol*. 2018;30:93–102.

4 Principles of Immunological Techniques

INTRODUCTION

Antigen–antibody reactions occur when an antibody binds to the antigen that matches its structure. These interactions are precise, as an antigen will react only with its own antibody or with one that recognizes a closely related antigen. In the body, the same reactions support immune defense by helping neutralize or remove microbes. However, they can also contribute to tissue injury, especially in hypersensitivity reactions and autoimmune conditions, where the immune system responds inappropriately to certain targets.

The antigen–antibody reactions serve several important functions. Their strong specificity allows laboratories to diagnose infections by detecting antigens or antibodies in clinical samples. This reaction is the basis for all important immunological tests employed for the detection and/or quantification of either antigen or antibody in infectious diseases.

ANTIGEN–ANTIBODY REACTION KINETICS

When an antigen interacts with the corresponding antibody, it sets the stage for an antigen–antibody reaction. Antigen–antibody reaction is a reversible chemical reaction, as shown here:

$$\text{antigen (Ag)} + \text{antibody (Ab)} \leftrightarrow \text{antigen} - \text{antibody (Ag-Ab) complex.}$$

This reaction results in the Ag-Ab complex that is held together by physical bonds. These include ionic bond, hydrogen bond, Van der Waals bond, and hydrophobic bonds, while covalent bonds are never formed. These forces are weak and act only for very brief moments, often lasting less than a second. A stable antigen–antibody bond forms only when many such interactions occur together. The epitope on the antigen aligns with the paratope on the antibody. Only a small group of amino acids on each molecule participates in this contact, and the interacting surface spans only a few nanometers. For the two sites to engage, they must be close enough to attract each other across about 3 nm.

The antigen and antibody molecules are attracted by forces, such as ionic and hydrophobic bonds. These attractive forces gradually overcome the hydration energies of the two molecules, with resultant expulsion of water molecules, so that the movement of the epitope and paratope toward each other is facilitated. Upon expulsion of interstitial water of hydration between epitope and paratope, a strong interfacial bond ultimately arises, which is mainly caused by Van der Waals forces (Van Oss, 1995).

The typical antigen–antibody reaction follows the law of mass action. According to it, in the earlier mentioned reaction: $\frac{[\text{Ag}-\text{Ab Complex}]}{[\text{Ag}][\text{Ab}]} = \frac{K_a}{K_d} = K_{eq}$, where K_a is association rate constant, K_d is dissociation rate constant, and K_{eq} is the equilibrium constant. This has significance in defining the affinity of an antibody mentioned below.

Affinity and Avidity

Affinity signifies the intensity of attraction between the molecules of antigen and antibody. In other words, it refers to the strength of the individual bond between the paratope and epitope contact residues. K_{eq} is inversely proportional to affinity, so the lower the K_{eq} value, the higher the affinity of the antibody.

DOI: 10.1201/9781003644224-5

High-affinity interactions are characterized by low K_{eq}, rapid recognition, and strong stability of the formed complex. The term *affinity maturation* is used to describe the gradual increase in the K_{eq} of antibodies produced in the body during immune response. Thus, early antibodies have equilibrium constants of around 10^5–10^6, while those produced after a few weeks or months show 100-fold greater affinities. In practical terms, when the K_{eq} value is 10^{-4} to 10^{-6}, the antibody sensitivity is micromolar; when it is 10^{-7} to 10^{-9}, the antibody sensitivity is nanomolar; and when the value is 10^{-13} to 10^{-15}, its antibody sensitivity is in femtomolar concentration. When an antigen is encountered for the first time, the affinity of the antibodies produced is low. Once the body is familiar with the antigen, the immune response matures and high-affinity antibodies are produced.

Avidity, on the other hand, is the functional affinity. It is the total binding strength of the antigen–antibody complex that is the sum of affinity, valency of antigen and antibody, and the structural arrangement of the epitope and paratope. Antibodies may be bivalent with two binding sites (IgG), quadrivalent (secretory IgA), or decavalent with 10 binding sites (IgM). Similarly, antigens that can bind to more than one antibody are multivalent. Avidity increases with the duration of an infection, and avidity assays are performed in toxoplasmosis and cytomegalovirus infections to find the relative time of infection.

Sensitivity, Specificity, and Predictive Values

Sensitivity, specificity, and predictive values are the parameters commonly used to discover the usefulness and accuracy of an immunological assay. It is imperative that large sample sizes are used while calculating these indices to arrive at a meaningful conclusion.

Sensitivity

Sensitivity is the frequency with which a test gives a true-positive result. It is the proportion of samples that are positive for the infection for which the test is done. In such cases a negative result rules out an infection. Sensitivity can be of two types: analytical and epidemiologic.

Analytic sensitivity of test is its potential to detect very small concentrations of the antigen or antibody. It is most useful in patient populations with a high incidence of disease who are seroconverting. It is dependent on various factors such as the test technique, the amount of antigen/antibody used, and the detector system. The best method to find analytical sensitivity is to use the seroconversion panel by repeated testing of an individual who has recently been infected at 1–2-week intervals.

Epidemiologic sensitivity of a test indicates the potential of the test to correctly identify the maximum number of individuals who have a definite infection. It is also called clinical sensitivity. Thus, these persons have high levels of the antigen or antibody being detected. It may not be true that an immunologic test can always detect these infections either due to the absence of a necessary antigen/antibody in the test format or inability to detect all genotypes. It can be assessed by testing all persons having an established infection:

$$\text{Sensitivity} = \frac{\text{True Positives}}{\text{True Positives} + \text{False Negatives}} \times 100$$

Table 4.1 provides the sensitivity of some routinely used important immunological techniques.

Specificity

Specificity is the frequency with which a test detects a noninfected patient so that no false-positive results are generated.

$$\text{Specificity} = \frac{\text{True Negatives}}{\text{False Positives} + \text{True Negatives}} \times 100$$

The dictum is that for the purposes of screening, a test with high sensitivity is desirable, whereas for confirmation, a test with high specificity should be used.

TABLE 4.1
Sensitivity of Common Immunodiagnostic Tests*

Test	Sensitivity (μg antibody/ml)
Precipitation tests	24–160
Agglutination tests	
Direct	0.4
Passive	0.08
Radioimmunoassay	0.0008–0.008
Enzyme-linked immunosorbent assay (ELISA)	0.0008–0.008
Immunofluorescence	8.0

* Adapted from Madigan MT, Martinko JM, Parker J, editors. *Brock Biology of Microorganisms*. 9th ed. Prentice Hall; 2000.

Predictive Value

The efficacy of a test is determined by a third parameter, namely prevalence of the disease. Since this parameter varies due to geographical locations, predictive values can also differ from one setting to another for the same diagnostic test. The predictive value of a positive test is the probability that a positive test is a true positive, and the patient has the disease, while for a negative test it reflects a person free from the disease. Thus, while choosing a particular test at a particular location, knowledge of the local prevalence of the infection is necessary. If the disease prevalence is low, a test with excellent specificity is preferred, whereas for high prevalence or high-risk group of patients, a highly sensitive test is needed.

$$\text{Positive Predictive Value} = \frac{\text{True Positives}}{\text{True Positives} + \text{False Positives}} \times 100$$

$$\text{Negative Predictive Value} = \frac{\text{True Negatives}}{\text{True Negatives} + \text{False Negatives}} \times 100$$

Nature of Antigen for Antibody Detection

The great majority of immunological tests rely on the detection of antibodies in the patient's blood or bodily fluids. These antibodies are detected against a specific antigen incorporated in the test format. In parasitic infections, a wide variety of antigens have been used in various formats, with varying degrees of success.

These antigens can be broadly classified as nonspecific antigens, whole or section of a parasite, somatic antigens (crude or purified extracts), excretory–secretory antigens(exoantigens), recombinant antigens, and hybridization antigens.

Nonspecific Antigens

During the previous century, several antigens derived from bacteria were studied that cross-reacted with parasite antigens. This led to the development of immunological tests for some parasites using these bacterial antigens. One such example was *Salmonella* Typhi, which reacted with an anti-trichinella serum in an agglutination test. Other studies have shown cross-reactivity between *Ascaris* and *Clostridium* and pneumococcus. Antigens from mycobacteria cross-react in *Leishmania* serologic tests. This was the basis of the now obsolete complement fixation test using Witebsky, Klingenstein, Kuhn (WKK) antigen prepared from human tubercle bacilli or using Kedrowsky's acid-fast bacilli.

Whole or Section of a Parasite

Schistosoma cercariae from snails or sections of worms recovered from experimental animals, frozen sections of adult filarial worms from dogs or cotton rat, blood films from humans or primates infected with malaria, and protozoa grown in culture like promastigotes of *Leishmania*, among others, have been primarily used in the past for direct fluorescent antibody test using human serum and are relevant even now. The major drawback, however, is the availability of the parasite in bulk, particularly for helminths.

Somatic Antigens (Crude or Purified Extracts)

Different developmental stages of a parasite can serve as antigens. In practice, stages that are easier to obtain in large amounts are used more often. However, these stages may show differences in their antigenic properties.

Once the parasite has been isolated in measurable amounts either through *in vitro* culture or cultivation in animals or tissue cultures, it is washed and stored by various means such as frozen state, lyophilization, or in glycerol.

Extraction of the antigen is carried out first by disruption of the parasite. Important physical methods used for disruption include grinding in a homogenizer, ultrasound disruption, repeated freezing and thawing, or osmotic shock. Homogenization is mainly used for parasitic worms, while ultrasound or freezing–thawing is employed for protozoa. Enzymes, lipid solvents, or detergents are some of the chemical methods that are also employed for disruption. Removal of lipid fraction either before or after the disruption step helps minimize nonspecific reactivity. Lipid solvents such as ethanol, ether, or acetone, or by detergents like Triton X-100 are also used to dissolve the protein. Other, smaller molecules can be removed by dialysis or ultrafiltration.

The goal is to obtain a purified extract that contains mostly the target protein antigen. Several methods are available to achieve this. Some of the simpler methods employed which produce products of variable purity are dialysis, ultracentrifugation and gradient centrifugation, various electrophoretic techniques (like isoelectric focusing, SDS-PAGE, etc.), ammonium sulfate precipitation, and salting out using sodium chloride.

To improve purity of the antigen, column chromatographic techniques are most useful. Multiple chromatographic formats are available, including ion-exchange chromatography and size exclusion chromatography, among others. Among these methods, affinity chromatography can produce a highly purified, target-specific antigen. This is possible only when a monoclonal antibody that specifically binds the antigen is available. Polysaccharide antigens are best obtained by affinity chromatography, although other techniques such as heating or extraction with phenol or alkali may be attempted.

Excretory–Secretory Antigens

Many parasitic helminths excrete or secrete various molecules inside the host to which immune response is elicited. These molecules are called excretory–secretory antigens. These excretory–secretory products (ESP) are part of the cuticular or tegumental surfaces of the parasite or in other cases secreted by specialized organs of the parasite. These products are shed in the surrounding environment and hence sometimes referred to as *exoantigens*.

To obtain the ESPs, the parasites are incubated or grown in a suitable liquid medium followed by centrifugation and then using the supernatant to isolate the ESP. Purification and concentration can be done by dialysis and ultrafiltration. Chemically, ESP is a mixture of glycoproteins, proteins, glycolipids, and polysaccharides, many of which are antigenic in nature.

Although studied extensively in helminths, ESPs have also been identified in protozoan parasites like *Toxoplasma*, *Giardia*, and *Entamoeba histolytica*. The ESPs have been used in several immunologic tests as the target antigen.

Recombinant Antigens

Recombinant antigens provide an excellent alternative to crude antigens, which have low sensitivity and specificity in antibody detection assays, and which require cultivation of the parasite.

The genome sequencing of organisms has opened a new avenue for identifying genus-or species-specific genes coding for unique proteins that can be used in immunodiagnosis (Figure 4.1). Previously it was thought that sensitivity of diagnostic methods using recombinant antigen is low; however, screening of the proteins in many parasites has resulted in identification of protein antigens that confer not only high specificity but improved sensitivity as well.

Both prokaryotic and eukaryotic expression systems have been used to generate recombinant parasitic antigens and have found their way into many commercial kits. It has been found that the success rates for obtaining highly immunogenic antigens for protozoan parasites is low if the *Escherichia coli* expression system is used, but that may not always be true. A wide range of eukaryotic expression systems are available and used extensively for production of both helminthic and protozoan proteins. These include yeast, mammalian cells, and baculovirus expression system in insect cells, among others.

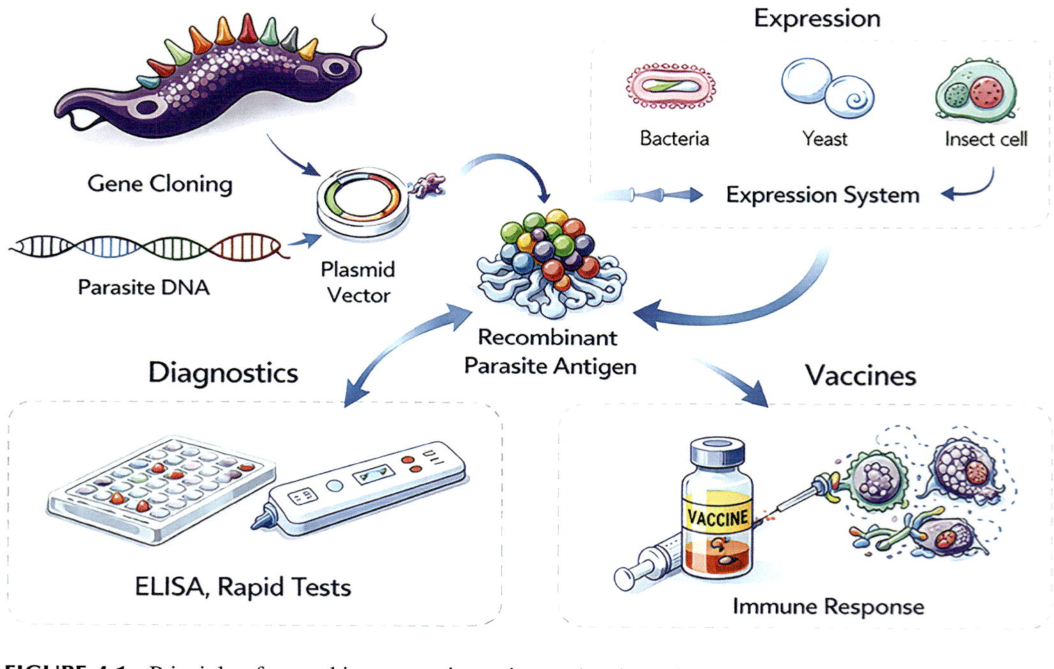

FIGURE 4.1 Principle of recombinant parasite antigens. A schematic representation of the production and application of recombinant antigens derived from parasites. Genes encoding immunologically relevant parasite antigens are isolated from parasite genomic or complementary DNA and cloned into suitable plasmid vectors. These recombinant constructs are expressed in heterologous expression systems, such as bacteria, yeast, or insect cells, to generate purified recombinant parasite antigens. The resulting antigens retain key immunogenic epitopes and are employed in immunodiagnostic platforms, including ELISA and rapid diagnostic tests, for sensitive and specific detection of parasite infections. In addition, recombinant parasite antigens serve as defined immunogens in vaccine development, where they elicit targeted host immune responses without the risks associated with whole-parasite preparations.

(The figure is copyright free and generated using an AI-assisted graphic illustrator.)

Recombinant antigen-based immunodiagnostic kits are commercially available for important infections like toxoplasmosis, leishmaniasis, malaria, giardiasis, amoebiasis, lymphatic filariasis, and schistosomiasis.

Hybridization Antigens

This type of antigen has been described for *Trypanosoma cruzi*, *Leishmania mexicana*, *Echinococcus* spp., and *Fasciola hepatica* where viable parasite cells were hybridized (fused) with mouse myeloma cells. Epimastigotes and promastigotes of the protozoan parasites and either the germinal cells or whole female parasite were fused with the myeloma cells, and the hybridomas were found to express the parasite antigen.

For detailed methodologies on the extraction, purification, and synthesis of parasitic antigens, readers are referred to the comprehensive work by Maizels et al. (1991).

Nature of Antibody for Antigen Detection

Two types of antibodies are used for detecting the antigens. The first antibody to be used was polyclonal antibody. It was subsequently followed by monoclonal antibody (mAb) following the monumental discovery of mAb synthesis by Georges Köhler and César Milstein in 1975.

Polyclonal Antibodies

Polyclonal antibodies are a mixture of heterogeneous antibodies produced by different B-cell clones in the body. Consequently, they can bind to many different epitopes of a single antigen.

The best source of polyclonal antibodies are rabbits, although other animals like guinea pigs, hamsters, chickens, sheep, goats, and camelids can also be used. Antibody production in chickens has the advantage of obtaining antibody through egg yolk without bleeding the animal.

To raise polyclonal antibodies, antigens are administered at least twice for best results. The second dose activates the memory cells that make class IgG antibodies against the antigen, resulting in a secondary antibody response. Affinity maturation leads to the production of antibodies that not only bind the antigen with much higher affinity but are also produced in greater quantities.

The addition of an adjuvant is necessary to boost antibody production. The best adjuvant is Freund's complete adjuvant, which is a mixture of mineral oil, a surfactant (mannide monooleate), and killed mycobacteria (*Mycobacteria tuberculosis* or *Mycobacteria butyricum*). The incomplete adjuvant does not contain mycobacteria. The adjuvant creates an antigenic depot effect resulting in slow and prolonged antigen release with prolongation of antibody production. The complete Freund's adjuvant is used for weakly immunogenic molecules, and only for the first dose. For subsequent doses, the incomplete adjuvant is used. Several other adjuvants are available commercially, such as Ribi Adjuvant Systems, Hunter's TiterMax®, Syntex Adjuvant Formulation (SAF)®, and Montanide ISA Adjuvants®.

Commercial polyclonal antibodies are also available, as obtaining them that way is more economical than generating them in the laboratories. The purity of the product can also be guaranteed.

Monoclonal Antibodies

Monoclonal antibodies have revolutionized not only the immunodiagnostics; they have wide-ranging applications as immunotherapeutic agents and for targeted drug delivery systems. Directed toward a single antigenic epitope, these antibodies are homogenous, highly specific, and can be commercially produced in unlimited quantities. The mAb technology has led to the development of highly specific diagnostic assays for parasites, which are free of the inherent drawback of cross-reactivity seen with polyclonal antibodies (Figure 4.2).

The mAb are produced either by the traditional mouse ascites method or by *in vitro* methods. The *in vitro* systems are based on cell culture and are of two types. The low-density cell culture utilizes a single-compartment system, while the high-density cell culture, which leads to higher

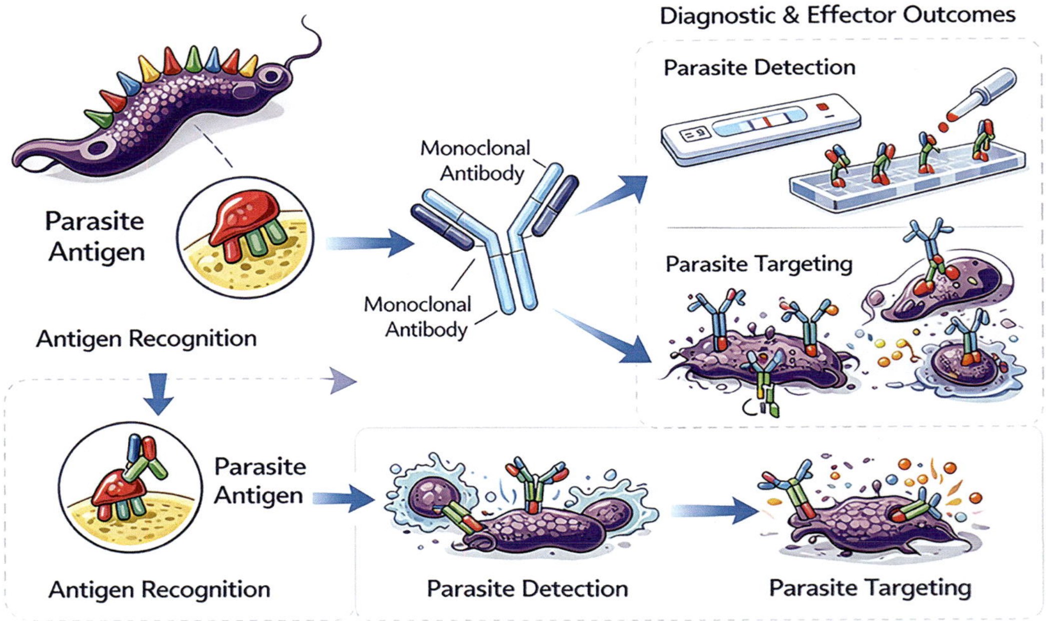

FIGURE 4.2 Principle of monoclonal antibodies (mAbs) in parasitic infections. The diagram illustrates the fundamental mechanism by which mAbs recognize and target parasites. Monoclonal antibodies with high epitope specificity selectively recognize parasite-specific antigens expressed on the surface or released during infection. Antigen–antibody binding forms the basis of immunodiagnostic applications, including rapid diagnostic tests and solid-phase immunoassays, enabling sensitive and specific parasite detection. In addition, antibody-coated parasites may be functionally neutralized or eliminated through immune effector mechanisms such as complement activation and antibody-dependent cellular cytotoxicity, resulting in parasite clearance.

(The figure is copyright free and generated using an AI-assisted graphic illustrator.)

concentration of Mab, needs double-compartment systems. For small-scale in-house production of a few grams, one of the low-density cell culture systems like culture flasks or roller bottles can be used. Bioreactors are employed for commercial production.

The mAb-based antigen detection kits are available as ELISA or rapid immunochromatographic tests for several parasites such as malaria parasite, *Cryptosporidium, Giardia*, *E. histolytica*, *Schistosoma*, *Taenia solium*, and *Wuchereria bancrofti*. The exquisite specificity of mAb makes them react with only one epitope, and hence only one site on the antigen molecule will be recognized. Hence, mAb are not a good choice for developing agglutination or precipitation type of tests, since both depend on the formation of multiple antigen–antibody interactions. The sensitivity of a test can be improved by using a cocktail of mAb.

CATEGORIZATION OF IMMUNOLOGICAL TESTS

In general, the various immunological tests may be categorized under three heads: screening tests, confirmatory tests, and research tests.

Screening Tests

These tests can detect most individuals who are truly infected. Hence, these tests should have a very high sensitivity index.

The majority of these screening tests detect IgG antibodies that persist for a long time and are produced in large volumes. However, these tests may miss infections during the window period, when antibodies have not yet appeared. They can also produce negative results early in the illness if only IgM is present and IgG has not yet developed. Mere detection of IgG is not helpful for many parasitic diseases endemic in a region, since low levels may be detected in many uninfected individuals due to subclinical or past infections. In general, IgG detection for diagnosis of parasitic diseases should be avoided in areas where the infection is endemic. For this reason, antibody detection, particularly IgG type, is not performed for diseases such as malaria, amoebiasis, filariasis, and the like in endemic areas of the world.

IgM detection in the screening test overcomes this problem to detect early infection, reactivation, or reinfection. Even in such scenarios, it must be kept in mind that IgM can persist for a long time in certain infections, such as toxoplasmosis.

Antigen detection tests may be the other screening alternative for parasitic infections. However, measurable amounts of antigen may be absent in the host in some infections, which may compromise the sensitivity of antigen detection tests. Currently, antigen detection tests as screening tests are widely used for malaria, amoebiasis, filaria, and schistosomiasis.

Confirmatory Tests

Confirmatory tests, also often referred to as the supplemental tests, are used to verify the positive results of a screening test. Such tests have high specificity to increase the predictive value. Sometimes they give indeterminate or negative results, since the sensitivity of these tests is somewhat lower. Such ambiguous test results may be followed up by other tests such as antigen detection or nucleic acid amplification tests. In most instances, these tests are not necessary, and a combination of clinical, radiological, and screening test results is adequate to arrive at a definitive diagnosis.

Research Tests

Many research laboratories devise in-house tests based on existing formats or using some novel technique. Some tests are also developed for infections for which no commercial test is available. Often, for various reasons, these tests remain confined to the laboratories without any commercial large-scale production. These tests may be utilized on a small scale by the clinicians of attached health care institutions or for referred samples from other hospitals. The reagents used may not be universally standardized, which can result in discrepant results at different laboratories.

Some highly specialized tests are also available for rare or neglected parasitic infections at DPDx (CDC); Liverpool School of Tropical Medicine; Institute of Tropical Medicine, Antwerp; Statens Serum Institute, Copenhagen; and a few other places. In India, PGI, Chandigarh, JIPMER, Pondicherry the RMRC, Dibrugarh, and many other institutes offer in-house tests for diagnosis of many parasitic infections.

These considerations can help clinicians and laboratory staff choose the most appropriate serological test, avoiding unnecessary procedures that offer little or no diagnostic value. While some advanced tests may appear promising, they do not always deliver the expected results.

CONCLUSION

Antigen–antibody binding forms the core of all immunologic diagnostic techniques. These reactions obey the law of mass action. The antigen–antibody complex is held together by weak, non-covalent forces. The equilibrium constant reflects the affinity between the two molecules. Avidity depends on affinity, the number of binding sites, and the spatial fit of the epitope and paratope. The performance of any diagnostic test relies on its sensitivity and specificity. The predictive value of a result changes with the local prevalence of the infection. Antibody detection tests often use crude or purified somatic antigens, excretory–secretory products, or recombinant antigens. Antigen detection assays may employ either polyclonal antibodies or monoclonal antibodies. Tests may serve as screening tools, confirmatory assays, or research methods. Screening tests need high sensitivity. Confirmatory tests require high specificity. Careful selection of the appropriate test is essential for an accurate interpretation.

FURTHER READINGS

Anonymous. Parasite antigens. *Bull World Health Organ.* 1975;52(3):237–249.

Fernández-Robledo JA, Vasta GR. Production of recombinant proteins from protozoan parasites. *Trends Parasitol.* 2010;26:244–254.

Foote J, Eisen HN. Kinetic and affinity limits on antibodies produced during immune response. *Proc Natl Acad Sci USA.* 1995;92:1254–1256. https://rsawa.research.ucla.edu/arc/polyclonal/. Accessed July 29, 2024

Kagan IG. *Characterization of Parasite Antigens*, 1967. Available at: https://iris.paho.org/bitstream/handle/10665.2/47517/RES6_2_2.pdf?sequence=1&isAllowed=y. Accessed July 24, 2024.

Karen DF, Warren JS. *Diagnostic Immunology*. Baltimore, MD: Williams and Wilkins; 1992a.

Maddison SE. Serodiagnosis of parasitic diseases. *Clin Microbiol Rev.* 1991;4:457–469.

Madigan MT, Martinko JM, Parker J., editors. *Brock Biology of Microorganisms*. 9th ed. Hoboken, NJ: Prentice Hall; 2000.

Maizels RM, Blaxter ML, Robertson BD, Selkirk ME. *Parasite Antigens, Parasite Genes: A Laboratory Manual for Molecular Parasitology*. Cambridge: Cambridge University Press; 1991.

National Research Council (US) Committee on Methods of Producing Monoclonal Antibodies. *Monoclonal Antibody Production. Vol. 5, Large-Scale Production of Monoclonal Antibodies*. Washington, DC: National Academies Press; 1999. Available at: https://www.ncbi.nlm.nih.gov/books/NBK100189/. Accessed July 29, 2024.

Parija SC. *Textbook of Medical Parasitology*. 4th ed. New Delhi: AIPD Publishers; 2013.

Parija SC. *Textbook of Microbiology and Immunology*. Singapore: Springer Nature; 2023a.

Van Oss CJ. Hydrophobic, hydrophilic, and other interactions in epitope-paratope binding. *Mol Immunol.* 1995;32:199–211.

5 Collection, Transport, and Preservation of Specimens

INTRODUCTION

Immunological diagnosis for parasitic infections relies on antibody or antigen detection most commonly in blood specimens. In special circumstances, however, other bodily fluids such as cerebrospinal fluid, stool, urine, and others may be tested. Specimen collection for immunodiagnosis of parasitic infections must follow strict timing and aseptic protocols. Containers should be sterile, properly labeled, and free from contaminants such as water or disinfectants. Blood is best collected during parasitemia, and stool samples should be fresh and collected on multiple days to improve diagnostic accuracy. Samples should be kept cool, unless otherwise instructed, to preserve antigenic and immune targets.

Specimens must be transported quickly in secure, leak-proof containers and maintained at appropriate temperatures, usually 2°C–8°C for most immunoassays. Transport media should only be used when necessary to avoid affecting antigen or antibody detection. In the laboratory, timely processing and validated storage conditions are essential to prevent degradation of immunologically relevant molecules.

LABORATORY TESTING OF SPECIMENS

Laboratory testing of samples for serological testing is typically divided into three main phases: pre-analytical, analytical, and post-analytical.

The pre-analytical phase includes all those procedures that are done before the performance of the test proper (Figure 5.1). It is a highly critical phase, since any error at this stage can adversely affect the accuracy of the test results. Studies have found that pre-analytical errors account for up to 70% of all mistakes made in laboratory diagnostics.

The analytical phase for immunoassays requires adherence to validated laboratory protocols. Reagents, controls, and calibrators are prepared according to manufacturer specifications, and instruments are calibrated and quality-checked before use. Serum samples, for example for serum-based immunoassays, are inspected for hemolysis or lipemia and tested in appropriate aliquots to avoid repeated freeze–thaw cycles. Assays are performed with positive and negative controls to monitor performance, and results are interpreted only within the assay's defined analytical range. Close attention to incubation times, temperatures, and detection conditions ensures accuracy and reproducibility in parasitic immunodiagnosis.

The post-analytical phase focuses on validating, reporting, and securely managing test results to ensure their clinical usefulness. Results of the assays are reviewed against quality control data. These are also checked for consistency with clinical information and known assay limitations. Any unexpected or borderline findings require repeat or confirmatory testing. Reports also include reference ranges and factors that may affect interpretation, such as hemolysis or delays in processing. Timely communication of critical results to clinicians is always essential. All records are documented and stored according to laboratory quality management standards.

Test Request and Patient Identification

Healthcare providers submit test requests containing relevant patient information and the tests to be performed. Accurate patient identification is essential to prevent errors, such as specimen mix-ups

DOI: 10.1201/9781003644224-6

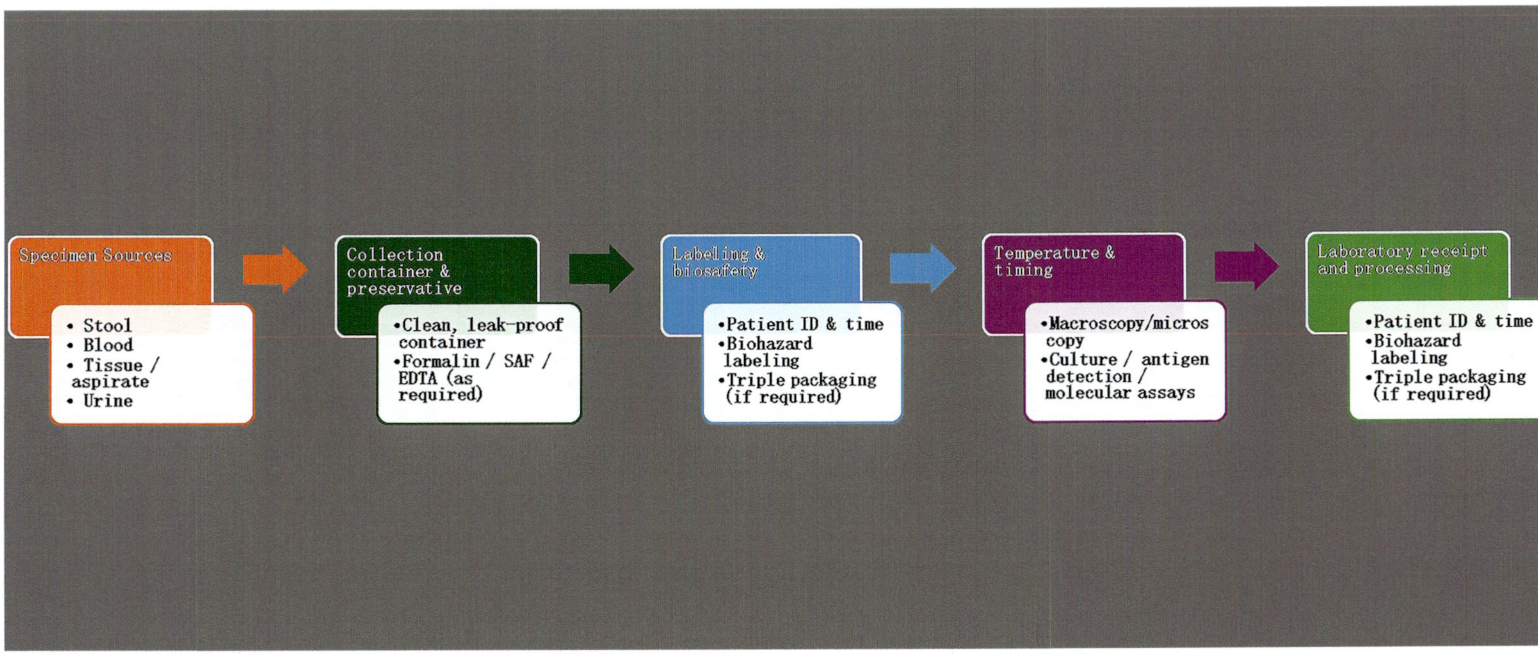

FIGURE 5.1 Key steps in specimen transport for parasitological diagnosis. Appropriate collection, preservation, labeling, temperature control, and timely transport are essential to maintain parasite morphology, viability, and molecular integrity.

(The figure is copyright free.)

and incorrect test results. With the widespread use of information technology, almost all accredited laboratories and health care institutions use some kind of software for management, and testing panels are built into these systems. This helps minimize human errors that were more common in the past, when requests were generated manually.

Laboratory professionals often take the initiative to guide clinicians in selecting the most appropriate diagnostic tests. This can be done in several ways, including continuing education seminars, laboratory publications of new tests being introduced and the existing test panels, or new uses of some older tests. With the current widespread use of information technology, the most effective way to help clinicians is to create requisition forms with clinical details in the hospital information system itself, which can facilitate the logical ordering of test sequences.

Specimen Collection

Improper collection of specimens can adversely affect the test results, and hence scrupulous attention needs to be given to this aspect. The correct specimen collected at the correct time is part of the diagnostic stewardship program. Care should be taken that the collected specimen maintains its integrity. For the great majority of immunological investigations, blood is the specimen collected. The next most common specimen is stool for antigen (coproantigen) detection. In addition, cerebrospinal fluid, pleural fluid, or urine may be collected in special circumstances.

Collection of Blood

The blood collection process may be conducted in accordance with the CLSI H03-A6: *Procedures for the Collection of Diagnostic Blood Specimens by Venipuncture; Approved Standard*, which provides detailed, standardized guidelines for safe and accurate venipuncture.

Although a phlebotomist is well versed with the techniques and related precautions needed for venipuncture, it is essential that laboratory personnel, nursing staff, and the clinicians are trained in drawing blood. Traditionally, blood is drawn into either a glass tube with no additives (for serum samples) or heparinized, ethylenediaminetetraacetic acid (EDTA), or sodium citrate tubes (for plasma specimen).

The practice of collecting blood in glass tubes is still being used in peripheral and remote health care settings, but it is being gradually replaced by a Vacutainer technique. Invented by Joseph Kleiner in 1949, these are sterile glass or plastic tubes with a vacuum inside and capped with colored rubber stoppers (for identification of different varieties of tubes for different investigations); the volume of vacuum determines the volume of blood that will be drawn using a particular tube. It is the exclusive trademark of Becton Dickinson, which first started the commercial production of such tubes. The types of capped tubes used for collection of serum from blood include red (no additive), gold (with clot activator and serum separating gel), and orange (with clot activator and serum separating gel), the last one being used for rapid serum separation. Plasma specimens may be collected in light green capped tubes containing heparin and plasma separator gel. Important details about collection, processing, and storage of blood/serum specimen are given in Table 5.1.

Collection of Stools

Fresh fecal samples are collected in a plastic container. Depending on the kit manufacturer's requirements, stool specimens are preserved in formalin or sodium acetate–acetic acid–formalin. However, concentrated, or polyvinyl alcohol (PVA)-fixed, stool specimens are not suitable for antigen detection. It is required to collect a minimum of 1–2 mL of liquid stool sample or 100–200 mg of solid stool.

Collection of Other Specimens

Cerebrospinal fluid and other bodily fluids are collected without any additives.

TABLE 5.1
Blood/Serum Specimens Used in Antigen-Based and Antibody-Based Immunodiagnostic Assays for Parasitic Infection: Collection, Processing, and Transport

Parameter	Antigen-based Assays	Antibody-based Assays
Purpose	Detect circulating parasite antigens, indicating active infection.	Detect host antibodies (IgM, IgG, IgA) indicating current or past exposure.
Collection requirements	Use sterile tubes; avoid hemolysis, as it can reduce antigen detectability.	Use sterile tubes; avoid hemolysis or lipemia to prevent assay interference.
Storage conditions	Store at 2°C–8°C short term; freeze at −20°C or −80°C for longer periods.	Store at 2°C–8°C short term; freeze at −20 °C or −80 °C for long-term stability.
Transport conditions	Maintain cold chain (2°C–8 °C); avoid temperature fluctuations.	Maintain cold chain (2°C–8 °C); protect from excessive heat.
Stability concerns	Antigens degrade quickly; avoid prolonged transit or delay in testing.	Antibodies are more stable, but avoid repeated freeze–thaw cycles.
Common sources of error	Contamination, hemolysis, improper temperature maintenance.	Hemolysis, lipemia, cross-reactive antibodies, improper storage.
Practical tips	Process and freeze aliquots early; use protease inhibitors if recommended.	Use paired sera when diagnosing recent infection; label aliquots clearly for retesting.

Specimen Transport

Whole blood can be stored at 2°C–8°C for up to 24 hours before the serum is separated, but it must not be frozen. If the serum is separated, it is stored at 2°C–8°C until shipment takes place, or for a maximum of 7 days. When kept for longer periods, serum samples are frozen at −20°C or lower and transported to the testing laboratory on ice packs. Repeated freezing and thawing of serum samples for IgM testing must be avoided, as it may have detrimental effects on the stability of IgM antibodies. As a rule, serum specimens should be shipped to the laboratory as soon as possible.

Stool specimens are transported in ambient temperature for short distances; otherwise, unpreserved stool should be frozen before transporting.

Specimen Processing

Serum

If a blood specimen is collected in a Vacutainer tube, the tube should be gently inverted no more than eight times to ensure proper mixing without causing hemolysis. The blood is then allowed to clot in an upright position for at least 30 minutes but not longer than 1 hour. This is followed by centrifugation for at least 15 min at 2,000–2,500 rpm. The separated serum is then transferred to a clean glass or plastic tube. It is usually stored at 2°C–8°C.

For blood collected without using a Vacutainer, it should be allowed to clot and then centrifuged at 3,000 rpm for 10 minutes to separate the serum. If there is no centrifuge, the blood can be kept in a refrigerator (2°C–8°C) until there is complete retraction of the clot from the serum, but no longer than 24 hours. The clot is loosened from the sides of the tube with an applicator stick, and the serum is carefully removed with a fine-bore pipette to avoid extracting red cells. The serum is then transferred to a sterile or clean vial.

It is to be noted that processing too quickly or delaying separation of the serum from the clot can increase the amounts of cellular products and degree of hemolysis, potentially affecting test results. Grossly hemolytic or cloudy serum (due to bacterial contamination or lipemia) is not recommended for immunological tests.

Stool

Normally, a 1:10 dilution of fresh or thawed patient stool specimen is prepared. For liquid stool, 100 μL of patient specimen is diluted with 1mL of antigen sample diluent. For solid stools, a small (pea-sized) portion of patient specimen is diluted with 1 mL of antigen sample diluent and mixed well. The diluted fecal sample is centrifuged at 3,000 rpm for 10–15 minutes. The supernatant is always used for the assay; however, the kit manufacturer's direction should be followed.

Specimen Storage

Serum samples are stored in tightly sealed containers in a non-frost-free refrigerator at 2°C–8°C for short-term storage. Frost-free freezers are avoided o minimize the freeze–thaw cycle. If serum is stored at 2°C–8°C for more than 2–3 days, bacterial or fungal contamination can occur. It may be avoided by using an additive like 0.05% sodium azide or 0.1% thimerosal.

For longer storage up to months or years, serum is aliquoted for cryopreservation at −20°C. A cryoprotectant such as 50% glycerol or ethylene glycol is added to the serum to prevent freeze–thaw damage. Storage at −80°C without additives can maintain the stability for 8–10 years or even up to 25 years, according to some studies. However, other studies have shown that there is degradation of IgM antibodies in serum stored at −20°C or −80°C. Freeze-drying or lyophilization with additives such as sucrose or mannitol has also been used for serum storage.

Stools are stored in a frozen state for extended periods, in either unpreserved state or with fixatives as mentioned earlier.

CONCLUSION

Proper immunodiagnosis of parasitic infections depends on rigorous specimen collection, timely transport, and controlled preservation. Clinicians should collect the appropriate specimen using sterile methods. Timing the collection is critical to optimize detection, such as obtaining blood during fever peaks or gathering stool on multiple days. Samples must be kept cool, transported quickly in secure containers, and handled without contamination. Prompt processing, validated storage conditions, and avoidance of repeated freeze–thaw cycles in the laboratory are essential to maintain antigen and antibody integrity. Accurate documentation of each step ensures reliable results and supports sound clinical decisions.

FURTHER READINGS

CLSI. *Procedures for the Collection of Diagnostic Blood Specimens by Venipuncture Approved Standard*, 6th ed. Wayne, PA: CLSI; 2007.

Parija SC. *Textbook of Medical Parasitology,Protozoology and Helminthology*, 4th ed. New Delhi: AIPD; 2023b.

Plebani M. Quality indicators to detect pre-analytical errors in laboratory testing. *Clin Biochem Rev.* 2012;33:85–88.

6 Quality Assurance in Immunological Tests

INTRODUCTION

Laboratory investigations that include immunodiagnostic procedures for parasitic and other infectious diseases have become an indispensable part of modern health systems. The ever-increasing laboratory services both in public and private sectors worldwide have necessitated the introduction of measures to check the results for their reliability comparable to those test results performed in other laboratories. This has led to the concept of quality assurance (QA) in laboratories to ensure generation of authentic results with the aim that no harm comes to the patient.

CONCEPTS IN QUALITY ASSURANCE

Quality: When a predetermined requirement of a user for a particular product or service is met with satisfaction, it is considered quality. The International Organization for Standardization code ISO 9000:2005 defines it as "the degree to which a set of inherent characteristics fulfils requirements." In serological and other investigations, a quality product or service can be defined as "the right result on the right specimen from the right patient that is accurate, timely and properly interpreted." This translates into an accurate report, which has been generated without delay, and the test results are similar as obtained from other laboratories. The results should also be properly communicated to the users (treating physicians) with appropriate interpretation.

Quality assurance: It is a comprehensive and wide-ranging concept that includes all the measures needed to generate an accurate and correct test result. It has been defined as "all the planned and systematic activities implemented within the quality system that can be demonstrated to provide confidence that a product or service will fulfil requirements for quality." The primary aim of QA is to prevent as many errors as possible and to accurately pinpoint the errors that do occur.

The key components of QA include assessment of the specimen quality; documentation of test validity; assessment of the performance of test procedure, reagents, instruments, and personnel; and review of test results and its clinical relevance (Figure 6.1). The term *good laboratory practice* (GLP) is sometimes used for QA. Thus, a QA program encompasses the following: quality assessment, quality control (QC), equipment monitoring, and audit (Figure 6.1).

Quality assessment: According to the World Health Organization (WHO), "assessment is the inspection of some part or the whole of the quality management system to show that the laboratory is meeting all regulatory, accreditation, and customer requirements." ISO uses "audit" in place of "assessment." In the audit or assessment, the performance of a laboratory is judged in two ways. It may be done by comparing with some predetermined standards or with other standard laboratories. The process may be performed by the laboratory staff itself or by some external agency.

Quality control: It is an integral part of quality management (QM) and deals with techniques and activities necessary to fulfill quality requirements. Analytical quality control (QC)

DOI: 10.1201/9781003644224-7

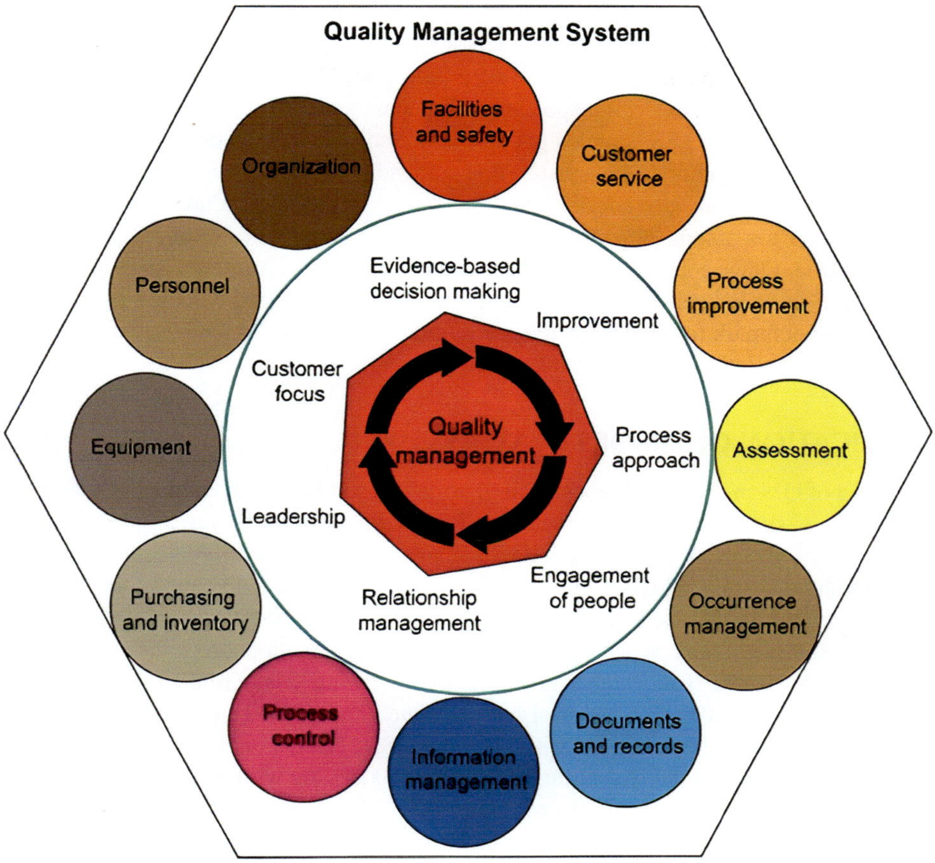

FIGURE 6.1 A comprehensive framework for the development and implementation of an effective quality management system in health care organizations, including 12 quality system essentials and 7 quality management principles.

(Photo courtesy of PathologyOutlines.com, contributed by Hai Duong Nguyen, MD, MSc. Used with permission.)

involves the equipment, reagents, calibration, and procedure reliability using standard operative procedure (SOP), right technique, expertise of operators, and internal and external QC.

Internal quality assurance (IQA)/control/internal audit: In IQA, all the activities starting from reception of the specimen to final dispatch of the reports are monitored. It is done at the laboratory level and may be done frequently, and necessary corrective measures can be instituted promptly. Each laboratory should have a clearly outlined policy about IQA, and national or international guidelines should be followed. This way it also prepares the laboratory for external quality assurance (EQA). In practice, 0.5%–1.0% of the total workload specimen is aliquoted and anonymized with a new request form where the patient's name is replaced with an IQA number, and this is resubmitted for testing. Any discrepancy between the original sample and the IQA sample is analyzed, and steps are taken to correct the anomaly.

External quality assurance scheme (EQAS)/control/external audit: EQA helps compare results among laboratories and detection systems. As the name suggests, assessment is done by an external agency or reference laboratory.

The most prevalent type of EQAS is proficiency testing. Here, the external agency sends a set of specimens for specific tests, the results of which are sent back to the agency for analysis and peer group comparison. The outcome of the analysis is sent to the laboratory with the desired result so that the participating laboratory can inspect any discrepancy and take remedial measures. There is the possibility that these EQAS samples get special attention and may be tested separately, or by senior staff, or tested repeatedly. This must be discouraged, and these specimens should be treated in a "routine" manner without any special attention. For example, the Indian Academy of Tropical Parasitology (IATP) is among other professional bodies in India that provide EQAS for parasitic infections in India. Other countries have their own EQA services. Another type of EQAS is a site visit by national or state agencies for inspection, certification, and accreditation of the laboratory. The ISO is the benchmark for all such activities, and the deficiencies of the laboratory are identified for the necessary corrective steps to be taken.

QUANTIFICATION OF ERRORS IN ASSAYS

The beginning of QM in clinical diagnostic laboratories started with clinical biochemistry. The basic principle used is statistical process control (SPC) with allowances for assay limitations and availability of QC materials. The mean and standard deviations (SD) obtained while testing the QC materials should reflect all types of methods, including reagents and calibrators, environmental factors, and operator variations. The selection of the correct SD will help in validating a test result.

The control sample(s) may be obtained from national or international repositories/laboratories or from commercial sources. This sample is tested in 20 separate assay runs along with the patient samples. The mean and SD of the control are calculated and plotted on Levey Jennings chart or Shewhart chart with the mean and ±1SD, ±2SD, and ±3SD. The QC samples demonstrate a normal, stochastic distribution of results. Accordingly, 68.3% of values are within ±1 SD of the target value, 95.5% are within ±2 SD, and 99.7% are within ± 3 SD. The acceptable limits are set at ± 3 SD to ensure high probability of error detection and low frequency of falsely rejected assay runs.

To determine the errors, Westgard rules are applied. It is a set of six rules of which the first three are warning rules while the other three are mandatory or alarm rules. Table 6.1 describes the details of the Westgard rules. These detect both systematic and random errors and specific performance limits. In summary, if the warning rules are violated, it necessitates review of the reagent quality,

TABLE 6.1
Westgard Rules and Their Interpretation

Westgard Rule	Criterion for Violation	Interpretation
1_{2SD}	If control value exceeds mean value by ±2 SD	It rules out a trend.
2_{2SD}	If two consecutive control values exceed mean value on the same side of mean by ±2 SD	This rule detects systematic errors.
4_{1SD}	If control value exceeds the same mean value by ±1 SD for four consecutive control values	This rule detects systematic error. The run need not be rejected if this rule is violated but requires recalibration or equipment maintenance.
1_{3SD}	If control value exceeds mean value by ± 3SD	This rule detects random error. Violation of this rule may also indicate systematic error. The results based on this assay are rejected.
R_{4SD}	If difference in SD between two control values exceeds ±4 SD	This is a range rule, and it detects random error. The results based on this assay are rejected.
10_X	If the last 10 consecutive values are on the same side of the mean	This rule detects a systematic error. Its violation often indicates the deterioration of assay reagents. The results based on this assay are rejected.

equipment performance, and calibration and test procedure. Any violation of alarm rules means results from patient samples for that assay run must be rejected. The details of the rules and an online Westgard QC calculator are also available at the website https://westgard.com/westgard-rules.html.

QUALITY MANAGEMENT OF SEROLOGICAL TESTS: PRACTICAL CONSIDERATIONS

Detection of antigen and antibody using immunological techniques is mostly done using commercial kits where detailed description of the methodologies, interpretation, and precautions to be taken in performance of the test is provided. In addition, these kits include both positive and negative controls, and in certain circumstances known standards of defined concentrations of antibody where quantitative detection is indicated. On the other hand, experimental/in-house test performance is commonly compared with a known commercial kit. In either of the situations, there is a need for internal quality controls to ensure the accuracy, sensitivity, and specificity of the test. Such controls can be obtained from commercial suppliers or reference laboratories.

One outstanding problem with serological tests for parasites is that such controls are available only for a handful of organisms, mostly protozoan parasites. In such situations, the laboratory must rely on its own confirmed positive and negative controls to validate a test. Here we describe a few important steps to be taken in any serological testing laboratory.

1. **Quality control:** The testing process needs to be monitored at every step for which the following measures are undertaken:
   a. Every test run should include appropriate controls.
   b. The controls must yield the desired result as per the manufacturer's specifications.
   c. The expiration date of the kits should be given due importance.
   d. Proper incubation time and temperature should be adhered to.
   e. The controls provided by the kit manufacturer are considered internal controls. Other controls that may be included in the test run should be considered as external controls. Such controls are necessary to ensure consistency of test performance, to check lot-to-lot variations in the supplied kits, and to indicate test performance in case of borderline results.
2. **Quality assurance:** There are multiple variables that can affect the results of any serological test. Hence, QA is a daily job and is an ongoing process that should be given due importance by the laboratory staff. Such activities include:
   a. *Specimen quality*: All specimens should be properly labeled with patient details. Nowadays, barcoded specimens are commonly used to prevent specimen mismatch between patients. Although blood/serum is the commonest specimen for serological tests, in special circumstances other bodily fluids like urine and cerebrospinal fluid are tested for antigen or antibody. These specimens should be tested without delay. In case a delay is anticipated, they should be kept at $4^{\circ}C$–$8^{\circ}C$ for 12–18 hrs. For long-term storage, serum samples should be kept at $-20^{\circ}C$ in a freezer. This is primarily done to prevent any microbial contamination. Serum samples that are grossly lipemic, hemorrhagic, or cloudy due to bacterial contamination should be rejected.
   b. *Documentation*: An up-to-date SOP is an essential component of any testing laboratory. The original source of the procedure along with subsequent changes must be documented in the SOP. In addition, records of day-to-day testing must be preserved and readily traceable, along with computer-generated data, for any future reference. Performance records of each test run and the QC procedures should be documented. A detailed discussion about SOP can be found later in the chapter.

c. *Laboratory personnel*: The laboratory technical staff as well as the supervisory personnel should be fully conversant with the various test procedures and the necessary precautions to be taken while performing the tests. A continuing education program should be conducted annually or semiannually to update the knowledge of the laboratory staff. Any knowledge gap in the staff should be addressed by the laboratory manager.

d. *Interpretation and transcription of results*: The interpretation of results must be done as per the kit instructions. However, for tests such as agglutination assays or complement fixation test, subjective reading of the results needs to be done by the performer of the test. In such situations, the results should be read by a second observer for corroboration. Many point-of-care rapid tests are read by the operator, but presently economical optical readers are available that can remove the subjective bias in interpretation of results.

Transcription of results signifies the dissemination of results to the end user, most commonly the treating physician. Transcription errors occur during transfer of results from test readout to hard copy and further to the computer in the hospital information system. In many small laboratories, hard copies of the test result aré handed over to the patients or the doctors. To minimize these errors, another technical staff should check the results, and the laboratory manager/supervisor should double check it before releasing the final report, which needs to be authorized by a qualified person.

3. **Quality assessment/audit:** This includes the following components:
   a. *IQA/Internal audit*: For IQA, the reference controls should be tested along with the routine samples, and the error should be identified using the aforementioned Westgard rule. In the event an error is detected, it should be investigated to pinpoint its origin. It may be with the personnel performing the test, with the test procedure or test kit, due to instrumentation errors, or a combination of such factors. A violation of 4_{1SD} rule points to equipment problem, which should be addressed accordingly (see later in the chapter).
   b. *EQA/External audit*: The relevance and importance of EQA have already been described earlier. It audits the laboratory organization and QA program. It provides a means to the laboratory supervisor to ensure that the laboratory's internal quality system is working optimally or not as assessed by an external agency. The proficiency panel of sera supplied by the EQA provider is tested without any bias. The outcome of the EQA must be discussed with all laboratory staff to correct any error that might have been flagged. In India, proficiency testing for parasitic infections is done by the EQAS program conducted by the IATP. The IATP panel testing is a three-tier system with microscopy, serology, and nucleic acid amplification tests, where the participating laboratory has the freedom to choose any one or more than one tier as per its scope of service. Similarly, the United Kingdom has its National External Quality Assessment Scheme (NEQAS) administered by the Public Health Laboratory Service (PHLS).
4. **Accreditation:** Accreditation is an audit of a laboratory's organizational structure and adherence to national/international standards. It can accredit the laboratory including all the tests, or individual assays can be accredited independently. The National Accreditation Board for Laboratories (NABL) is the national agency in India. Similarly, in the United States it is done by College of American Pathologists, in the United Kingdom by Clinical Pathology Accreditation, and by National Association of Testing Authorities in Australia. In essence, they follow the ISO 15189, which is the international standard for medical laboratories.
5. **Batch-to-batch variation and reagent quality:** It is an established practice that there may be reagent variation in a different lot of reagents or diagnostic kits. This may occur on the manufacturer's end or arise due to improper transportation and cold chain maintenance, not only for imported kits and reagents but also for local products in tropical countries. Variations in new lots must be detected before they are used for patient testing. The laboratory should decide about the following:

- Concentration of the reagents necessary for patient testing
- Number of samples to be evaluated for each concentration
- Relevant rejection limits.

For qualitative assays such as agglutination tests or qualitative ELISA, a new batch of reagent or kit is validated by QC confirmation. False-positive results are also monitored. For quantitative assays, such as IgG concentration by ELISA, a panel of three serum samples should be evaluated. These are the negative, weak positive, and strong positive samples tested in the previous batch, which are tested in triplicate. A drift between the reagent lots can be ascertained by monitoring the false positive or non-confirmatory results. These are the samples that tested positive in the assay but negative in a confirmatory test. If the reference change value is more than 50% for any two results, the new lot should be rejected, since there will be a clinically significant bias. For details on reagent lot variation assessment, the reader can consult the CLSI.EP26 recommendations.

6. **Standard operating procedure (SOP):** The SOP is the most important document in a laboratory. It helps produce reliable and reproducible results and describes in detail each of the test procedures performed in the laboratory. It must be exhaustive and accurate and should conform to the laboratory policy. The SOP should be readily available at each workstation and should be periodically reviewed (preferably every year) and updated as per the laboratory requirements. This SOP must be strictly adhered to by all laboratory personnel. It should be written by persons performing the test and then modified and edited by the supervisor before releasing the final version.

An SOP has two distinct sections. The first details the laboratory structure and describes the SOP; the second deals with the procedures.

The following is the format of the first section, in sequential order:

1. Name of the organization and the laboratory or section using it (parasitic serology).
2. Name of the SOP.
3. Unique identification code and version number.
4. Issue date.
5. Name of the person preparing it.
6. Name of the person authorizing it.
7. Laboratory personnel organogram indicating the line of authority.
8. Laboratory safety features including emergency measures.
9. Page number with reference to total page numbers (e.g., 54 of 150).

The procedure section of the SOP includes the following:

a. Name of the test, introductory note, scope of the test, and definitions.
b. Specimen collection: It includes patient selection, procedure for collection, transportation, storage, sample rejection criteria, and steps to be taken if sample is rejected.
c. Details of procedure:
    - Title
    - Principle of the test
    - Sample preparation/processing (e.g., separation of serum procedure)
    - Reagents needed: source, dilutions, storage
    - Instrument needed
    - Standards and controls: source, dilutions, storage
    - Directions for test performance
    - Results: Its interpretation. If the results are read visually, how to differentiate positive and negative results. In instrumental readings, the cut-off value needs to be calculated before interpretation
    - Troubleshooting guide: source of common errors and its remedy.

   d. Quality control: The policy of the laboratory regarding time and frequency of running QC steps (e.g., when a new batch of reagents are being used). Criteria for accepting the QC results and measures to be taken if there is deviation from the acceptable results must be clearly stated in the SOP.
   e. Reporting of result: The hierarchy to be followed for releasing the results must be clearly spelled out.
   f. References: Protocol or guidelines, books, and instruction kits are to be mentioned for ready reference.

7. **Instruments and accessories:** Instrumentation is an integral part of any laboratory, and advanced reference laboratories may have multiple sophisticated instruments. While going for the purchase of any instrument, whether a basic one like a centrifuge or a high-end like multimode reader, certain considerations should be kept in mind.

   These include the following:

   a. The specifications of the instrument should meet the scope of the laboratory and the purpose, and it should be cost effective.
   b. The instrument should conform to the infrastructure of the laboratory.
   c. Servicing of the instrument and any repair work in case of breakdown should be prompt, and the concerned engineers or technical staff should be easily accessible.
   d. An annual maintenance contract with the supplier is economical in the long run.
   e. The supplier should train the laboratory staff in running of the instrument and supply the operation and maintenance manual along with a troubleshooting guide.
   f. The running cost should be reasonable, and operation of the instrument should not be a complex one.
   g. It is preferable to procure an instrument that has an "open system," meaning thereby that it can be operated with reagents and controls from diverse manufacturers. "Closed system" instruments can only be operated by reagents supplied by the same manufacturer.
   h. It is always a good idea to have a "reagent rental" arrangement whereby the laboratory pays only for the reagents or kits from the supplier without paying for the very expensive instrument. Such an arrangement works well if the quantum of a particular test performed is quite high.

   Once an instrument is procured, careful attention should be given to its day-to-day operation. Some of the important ones include the following:

   a. Temperatures of incubators, refrigerators, deep freezers, and water baths must be monitored daily.
   b. Rheostat control of a centrifuge should be checked annually or semiannually with a tachometer to ensure proper gravitation force. It can be done by the biomedical engineer.
   c. Calibration is a process of quantitatively measuring an instrument's performance to check its proper functioning at all measuring limits. It is a must for such instruments as spectrophotometers, automatic washers, automatic pipettes, and nephelometers. Calibration should be done by the experts approved by the supplier and is an important part of an annual maintenance contract.
   d. Logbooks must be maintained for each instrument recording the use of the instrument, including the name of the test run, time when operated, and the name of the operator. If any problem is detected with the instrument, it should be documented, and the troubleshooting guide or instrument manual should be consulted. The laboratory manager needs to be informed about the problem and the outcome of any remedial steps taken by the operator.

Apart from the instrument, several accessory pieces of equipment are commonly used in a laboratory. The glassware used in a serology laboratory should be clean and free of detergents. Damaged or cracked glassware should be discarded. Volumetric glass pipettes should be checked internally for proper calibration. Semiautomated pipettes can be checked to see whether they are dispensing the correct volume of the fluid. Similarly, pipette tips should be clean and should not be reused.

Challenges in Quality Assurance in Serological Testing for Parasites

In contrast to clinical chemistry, where most of the analytes are small homogenous molecules, serological assays detect antigens or antibodies that are more complex molecules, with a high degree of heterogeneity. These elements are called Class B quantities, which measure functional biological activity in the body.

In contrast to clinical chemistry, serological tests face the following challenges: lack of international standards; lack of certified reference method or gold standard; antigens are of various types and heterogenous; most of the tests are qualitative with results expressed only as positive or negative; and nonlinear dose–response curve. Nonavailability of standard sera and defined antigens of parasites is a serious problem associated with serological tests, which leads to difficulties in standardizing a test when it is developed.

Standardized antigen or reference sera are available for a few animals' parasitic diseases from the World Organization of Animal Health. Quantification of antibodies is another problem area, where concentration is expressed in International Units, whereas in many commercial kits the information about reference sera used for calibration is missing. Similarly, there is an acute dearth of reference laboratories where samples can be sent for confirmation. The few existing ones are located in Europe or the United States, and the tariffs are prohibitively expensive for laboratories located in Southeast Asia, Africa, or South and Central America.

CONCLUSION

Maintenance of quality standards in serological laboratories depends on multiple factors, including laboratory staff, infrastructure, and quality of reagents. The requirements may vary depending on the level of laboratory (small, intermediate, large, and/or reference laboratory), but the basic tenets of QA hold true at all levels. All such laboratories should adhere to the principles of GLP and pay particular attention to QC. One of the most frustrating challenges facing a serological laboratory specializing in parasitic diseases is the lack of suitable reference laboratories that can be important sources of standardized antigens and antibodies. This shortcoming seriously affects the QC parameters that are essential components of any serological assay. In this regard, information obtained from the ever-expanding horizons of genomics and proteomics can be applied to the production of recombinant antigens and monoclonal antibodies, which can go a long way to solving this age-old problem.

FURTHER READINGS

Badrick T, Fortun M, Vayanos Z, Bernard M, Dufour P, Souied L, Giannoli JM. Quality control for serological testing. *Clin Chim Acta.* 2025;564:119905.

CLSI.EP26. *User Evaluation of Acceptability of a Reagent Lot Change.* Malvern, PA: CLSI; 2022.

Dimech W. The standardization and control of serology and nucleic acid testing for infectious diseases. *Clin Microbiol Rev.* 2021;34(4):e0003521.

Public Health England. (2013a). *Quality Assurance in the Diagnostic Virology and Serology Laboratory*. UK Standards for Microbiology Investigations. Q2(6):2. Available at: https://www.hpa.org.uk/SMI/pdf

World Health Organization. Regional Office for South-East Asia. *Quality Assurance in Bacteriology and Immunology*. 2nd ed. WHO Regional Office for South-East Asia; 2003. Available at: https://terrance.who.int/mediacentre/data/ebola/training-packages/LQMS/11_cd_rom_quality_assurance_in_bacteriology_and_immunology_2002.pdf

World Health Organization. *Laboratory Quality Management System: Handbook*, 2011. Available at: https://iris.who.int/server/api/core/bitstreams/0772adf4-743a-48bb-843b-7ef4c2dc6896/content

Part II

Immunodiagnostics in Parasitic Infections

Unit 1

Antigen- and Antibody-Based Serological Tests

7 Carrier Particle-Based Agglutination Tests

INTRODUCTION

Many immunological tests are based on antigens or antibodies that have been covalently linked to or adsorbed on particulate carriers. Various particles have been used, but the ones employed most are erythrocytes, latex particles, and charcoal. This format is also known as passive or indirect agglutination in contrast to active or direct agglutination where no carrier particle is involved. Direct agglutination is used primarily for bacteria and blood group antigens and will not be discussed in this chapter.

PRINCIPLE OF CARRIER PARTICLE-BASED AGGLUTINATION TESTS

Inert particles can attach to antigen or antibody molecules with or without some form of pretreatment. The desired molecules are either adsorbed or covalently linked to the particles. Erythrocytes from sheep or rabbit or from human blood group O readily adsorb a variety of antigens and can be stored after treatment with formalin, paraformaldehyde, or glutaraldehyde. Mild treatment with tannic acid or other oxidizing agents helps in better antigen adsorption by the red blood cells.

Other particles such as latex, bentonite, or charcoal can also adsorb some antigens or antibodies, but latex particles with exposed carboxyl groups can better attach to the molecules through covalent linkage. When these particles are allowed to react with the patient's samples containing the corresponding antibody or antigen, there will be binding of the corresponding molecules, resulting in visible clumping of the particles (Figure 7.1).

INTERPRETATION OF THE TESTS

Agglutination tests are carried out on slides/cards/microtiter plates and are semiquantitative in nature. Interpretation is sometimes difficult because of the subjective nature of detecting the end point by the observer. Results are mentioned as titers or on the scale of 0 to 4+. Although specificity may remain unchanged, sensitivity varies, and IgM antibodies are 400–750 times more efficient in agglutinating property compared with IgG.

MECHANISM OF AGGLUTINATION

Agglutination is clumping of antigen/antibody-coated particles by the corresponding molecules, forming bridges. It occurs in two stages:

1. *Sensitization*: The first stage of sensitization involves attachment of antibody/antigen molecule to the corresponding antigen/antibody fixed on the particle. This is a reversible interaction and depends on the affinity or equilibrium constant of the antibody. The degree of associations depends on the antigen–antibody ratio and on various physical factors such as pH, temperature, ionic strength, time of interaction, and steric hindrance. The optimum pH used is 7.0, and optimum temperature for IgM detection is 4°C–22°C, while for warm reacting IgG it is 37°C. The ionic strength has a profound effect on agglutination reactions, and sodium chloride (NaCl) is the preferred milieu for the interaction to occur.

DOI: 10.1201/9781003644224-10

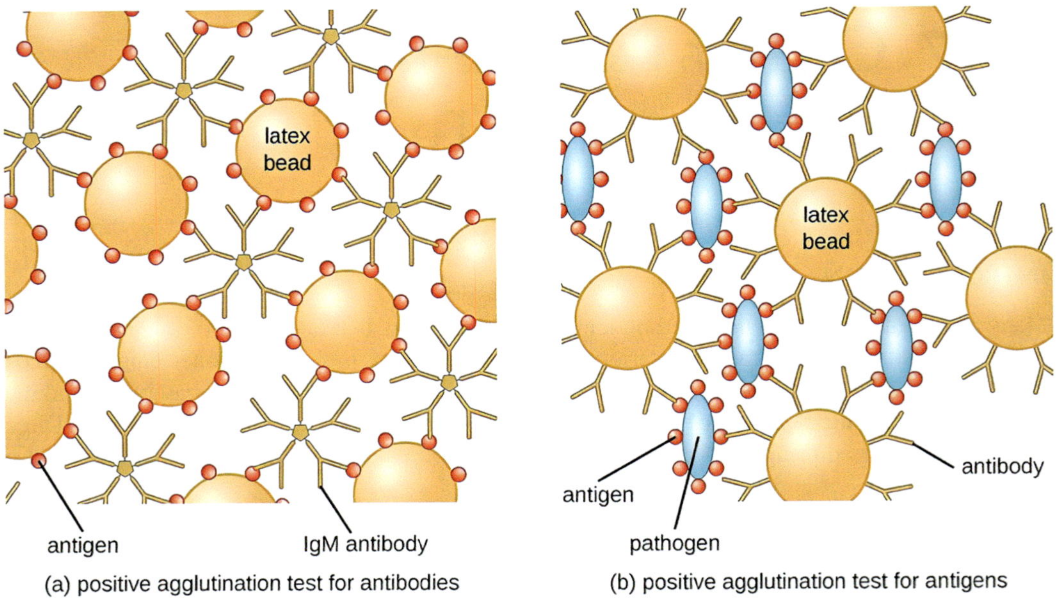

FIGURE 7.1 Lattices formed by carrier particles (latex beads). OpenStax Microbiology. Provided by OpenStax CNX. Available at: http://cnx.org/contents/e42bd376-624b-4c0f-972f-e0c57998e765@4.2.

The Na^+ and Cl^- ions help neutralize the opposite charges on the antigen and antibody molecules, thus facilitating a more efficient association between them. Use of low-ionic-strength saline enhances the association.

2. *Lattice formation*: Cross-linking of the coated particles and the analyte molecules occurs at this stage, with the formation of chemical bonds and lattice. It is a longer process compared with sensitization, but it is more stable.

Agglutination reactions may be enhanced by various techniques. These techniques help bring the particles closer and include centrifugation or mild proteolytic enzyme treatment to lower the zeta potential.

PARTICLE-BASED AGGLUTINATION TEST FORMATS FOR ANTIGEN DETECTION

While the preceding discussion pertained to latex agglutination tests for antigen and antibody detection, two specialized formats can be employed exclusively for antigen detection. These include reverse passive agglutination (RPA) using latex particles and reverse passive hemagglutination (RPHA) using red blood cells (RBCs) and co-agglutination (COA) in which *Staphylococcus aureus* is used as carrier particles.

Reverse passive agglutination: In this assay, antibody is bound to the carrier particle with the active sites facing outward. Either spontaneous absorption or some chemicals are used for attaching antigens. These particles react with the patient's sample containing the antigen when agglutination occurs. The sensitivity of these tests is high, and patient's serum, cerebrospinal fluid, or urine can be tested.

Co-agglutination test: *Staphylococcus aureus* Cowan 1 strain has a protein A on its surface, which can naturally adsorb F_C portion of antibody molecules, leaving the antigen-binding F_{ab} sites open. When these antibodies-coated organism reacts with the patient's sample containing the corresponding antigen, clumping of the organisms occurs. It shows higher stability than latex particles and is more resistant to any changes in ionic strength. Specificity is high, but sensitivity for detecting low concentrations of antigen may be lower.

Applications in Parasitic Diagnosis

The indirect hemagglutination assay (IHA) and the latex agglutination test (LAT) have been primarily used to develop antibody detection kits for the diagnosis of parasitic infections. Many academic and research laboratories have developed and evaluated such assays for diagnosis of various parasitic infections, with varying results.

The important parasitic infections for which such tests have been developed include amebiasis, giardiasis, malaria, babesiosis, toxoplasmosis, toxocariasis, echinococcosis, schistosomiasis, filariasis, strongyloidiasis, and many others. Only a few commercial kits, such as the toxoplasma latex agglutination test, are currently available on the market (Table 7.1).

The following sections outline the in-house development and design of representative antibody and antigen detection methods using IHA and LAT, as well as antigen detection through the co-agglutination test. It includes:

1. Designing the IHA test for *Echinococcus* serum antibody detection.
2. Designing a reverse passive hemagglutination (RPHA) test for *Cryptosporidium* antigen detection in human stool.
3. Designing the LAT for *Toxoplasma* serum antibody detection.
4. Designing the Co-A test for *Echinococcus* antigen detection in urine and serum (as developed by Parija and associates).

Designing and Optimizing the Indirect Hemagglutination Test for *Echinococcus* Serum Antibody Detection

As mentioned earlier, human/sheep/chick red blood cells are used as the carrier particle in this test. The RBCs are coated with the echinococcal antigen and reacted against patient's serum. Presence of antibody in the serum is detected by agglutination of the RBCs, which can be visualized macroscopically. This chapter describes the in-house design and optimization of the IHA for Echinococcus serum antibody detection in serum by Parija and Ananthakrishnan.

TABLE 7.1
Commercially Available Kits in Carrier Particle-Based Agglutination Tests

Test Format	Disease
Indirect hemagglutination	Amoebiasis
	Schistosomiasis
	Toxoplasmosis
	Echinococcosis
	Distomiasis
	Chagas' disease
Latex agglutination test	Toxoplasmosis
Card agglutination test	*Trypanosoma brucei gambiense* infection

Note: For the suppliers of these kits, please see Appendix B.

Reagents and Equipment Required

Reagents: hydatid antigen; Alsever solution; double aldehyde stabilized human/chick cells; phosphate-buffered saline (PBS) (pH 7.1, 7.2, and 6.4), PBS (pH 7.2) with 0.1% bovine serum albumin; tannic acid; pyruvic aldehyde; aqueous glutaraldehyde (2% v/v); Sorensen's phosphate buffer (pH 7.2); sodium chloride; sodium carbonate.

Equipment and labware: automated pipettes, U-bottom microtiter plates; centrifuges.

Design and Optimization of the IHA Test

Developing the IHA test can be divided into the following steps:

Preparation of Antigen

A wide range of antigens have been used in the development of IHA. The most common is the hydatid cyst fluid (HCF) antigen derived from sheep, cattle, or human hydatid cysts. Sensitivities of serological tests based on HCF are high, but their specificities vary.

Other antigens that have been used include antigen 5, antigen B, protoscoleces antigen, and whole larval antigen. Of these, HCF and Antigen B or Antigen 5 (native, synthetic, or recombinant form) have been used most frequently. For the preparation of individual native antigens 5 or B, please consult the study by Pagnozzi et al.

The following steps describe the preparation of HCF that is rich in both antigen 5 and antigen B:

- A hydatid cyst removed surgically from humans or sheep/cattle is the source of HCF.
- The intact, unruptured cyst is transported to the laboratory without formalin preservative.
- It is washed carefully with sterile PBS, and fluid is aspirated carefully in an aseptic manner with needle and syringe.
- The aspirated fluid is sterilized by filtration using a Seitz or membrane filter.
- Bacterial contamination should be checked by inoculation in the blood agar and MacConkey agar.
- The protein content is estimated using an autoanalyzer, followed by storage at −20°C in small aliquots of 1 mL.

RBCs from Group O individuals or from sheep or chick can be used.

- Sheep blood, human blood or chick blood is collected under aseptic conditions into an equal volume of sterile Alsever solution. It is labeled with the date of collection and stored at 4°C for no longer than 1 month.
- Alsever solution is prepared as follows: glucose, 2.05 g; sodium chloride, 0.42 g; sodium citrate dehydrates, 0.8 g; anhydrous citric acid, 0.055 g; and distilled water to 100 mL. It is sterilized at 121°C for 15 minutes. pH is adjusted to 6.1.
- 20 mL blood in an equal volume of Alsever solution is centrifuged at 10,000 rpm for 10 minutes. The supernatant, which should be clear, is removed.
- The sediment is washed with 20 mL PBS. For this, it is mixed by gently inverting the tube and centrifuged as described above, repeated three times.
- The RBC pellet from the final third wash is used to make a 10% solution in PBS (2 mL blood cells in 18 mL PBS).

Red blood cell stabilization is done by treatment with tannic acid or preferably by the double aldehyde stabilization (DAS) method. This step is important, since it makes the cell surface anionic by removing the free amino groups and thus making the cells more lipophilic. This in turn increases the affinity of the cells for the protein antigen through salt-like or nonionic forces.

Tannic acid method: This is a simple method used extensively in the past. A stock solution of tannic acid is prepared at a concentration of 1 mg/mL and stored at 4°C for no longer than 3 months. Before use, the stock tannic acid solution is diluted with PBS pH 7.1.

- A 1% suspension of sheep red cells is prepared in PBS from 10% RBCs.
- Equal volumes of the 1% RBCs are mixed with the tannic acid solution and incubated at 37°C for 15 minutes, after which it is washed once in PBS.
- It is then centrifuged, and the RBC sediment is used for further work.

Double aldehyde stabilization method: In recent years, the tannic acid treatment has been largely superseded by the DAS method. Treating RBCs with two aldehydes, specifically glutaraldehyde and formaldehyde, is considered more efficacious for several reasons. The RBCs become resistant to hemolysis and can be stored for up to 1 year at 4°C; there is increased agglutination property; and the aldehydes make the RBCs more anionic, resulting in firmer attachment of antigens. The following method, as described by Parija and Ananthakrishnan, can be followed:

1. Stabilization with pyruvic aldehyde:
   - A reaction mixture is prepared in the following sequence and then kept at 4°C: 1.7% NaCl (12 mL), 40% v/v aqueous pyruvic aldehyde (4 mL), 1.0% Na_2CO_3 (35 mL), Sorensen's phosphate buffer (pH 7.2) (7 mL).
   - To this, 10 mL 50% suspension of chick RBCs is added. The mixture is gently stirred for 30 minutes and stored for 24 hours at 4°C.
   - The next day, RBCs are washed thrice in PBS, and a 2.5% v/v suspension is prepared.
2. Stabilization with tannic acid followed by glutaraldehyde:
   - To a 2.5% suspension of the pyruvic aldehyde–stabilized RBCs, an equal volume of cold tannic acid, 1:25,000 in PBS, is added. Then the mixture is kept at 4°C for 30 minutes.
   - This is followed by three washes in PBS, and a 4% suspension of RBCs is made in PBS.
   - An equal volume of 2% v/v aqueous glutaraldehyde in PBS is then gently added and blended with a magnetic stirrer for 2 hours.
   - Finally, three washes with PBS are performed again and then a 50% solution of DAS RBCs is prepared and stored at 4°C.

Determination of Optimum Sensitizing Dose of the Antigen

The optimum sensitizing dose (OSD) of antigen is determined for the stabilized cells by checkerboard titration with known positive and negative sera. The OSD is the lowest titer of antigen that shows maximum agglutination.

- Doubling the dilution of antigen is done in PBS starting from 1:2 dilution to 1:256 in eight tubes.
- To each tube 0.1mL of 10% stabilized (DAS/tannic acid) chick RBCs in PBS is added, mixed thoroughly, and incubated at 37°C for 1 hour, with careful mixing every 15 minutes.
- The mixture is then centrifuged for 5 minutes at 4,000 rpm, and the supernatant is discarded.
- The RBCs are washed three times in 2mL PBS, and the supernatant is again discarded.
- The RBCs are resuspended in 990μL of PBS-BSA (bovine serum albumin) to produce the final concentration of 1% sensitized red cells.
- The solution can be stored at 4°C for 24–48 hours.

For the checkerboard titration, a 96-well microtiter plate is used.

- 50μL of PBS/BSA is dispensed into all wells of columns 1 to 12.
- Then 50μL of absorbed positive control serum is added to all wells in column 1.
- Using a multichannel pipette, 1:2 dilutions are made from column 1 to column 11 for each row, discarding the final volume. The final dilution of serum will then be 1:10 to 1:10,240 from column 1 to 11.
- In the next step, 25 μL of chick RBCs from each of the antigen dilution tubes is added into each well of the adjacent row from column 1 to column 12 (e.g., 25μL of cells from antigen dilution 1:20 is added to each well of row A, 25μL of cells from antigen dilution 1:40 is added to each well of row B, etc.). The RBCs in rows A to G are sensitized while row H contains non-sensitized red cells.
- The plate is gently shaken for mixing, covered with aluminum foil, and left on the bench at room temperature for 2 hours, then at 4°C overnight. Plates can be read at either time point.
- Negative wells (no agglutination) have an intact button at the bottom of the well. Positive wells (red cell agglutination) demonstrate a fine carpet of cells or appearing as a loose button with ragged or folded edges.
- The plates can be read with a reading mirror for microtiter plate with light transmitted from below.
- The OSD is the lowest titer of antigen that shows maximum agglutination. This titer is used for further sensitization of the RBCs. In the study mentioned above, the OSD was found at 1:16 dilution of the antigen.

Test Procedure

Specimens: Serum sample to be tested.

RBC sensitization with OSD of the antigen:

- All reagents are brought to room temperature before the experiment.
- A volume of 0.1 mL of DAS packed chick RBCs is washed twice with PBS, and 0.9 mL of OSD titer of the antigen is added.
- The cell plus antigen mixture is incubated at 50°C in a water bath for 5 minutes, followed by overnight storage at 4°C and then incubation at 50°C again for 5 minutes.
- The sensitized cells are washed thrice with PBS and stored in small volumes as a 10% suspension in PBS at 4°C until used.
- The sensitized chick DAS cells retain their sensitivity for up to 31 days after preparation when stored at 4°C.

Performing the test: The test is carried out in a U-bottom microtiter plate.

- The test serum is de-complemented by placing it in a water bath for 30 minutes at 56°C.
- Next, 25 μL of assay diluent (PBS-BSA, pH 7.2) is dispensed in each well of a row (12 wells), and 25 μL of serum is added to the 1st well of the row and then diluted until the 11th well, leaving the 12th well as serum-free control.
- Next, 25 μL of 1% sensitized RBCs are added to each well and the plates are agitated for 2 minutes. Results can be read after incubation at room temperature after 30 minutes.
- A reading mirror can be used to read the plates with light transmitted from below.

Result and interpretation: A smooth carpet of cells or loose aggregate with ragged edges signifies a positive test, while a compact button of cells denotes a negative test. The highest dilution of serum giving a positive reaction is taken as the reactive titer.

Quality assurance: Known hydatid antibody positive and negative controls should be run simultaneously as quality controls with every test. A well containing only sensitized RBCs without addition of serum is used as cell control.

Design and Optimization of a Reverse Passive Hemagglutination Test for *Cryptosporidium* Antigen Detection in Human Stool

A modification of the IHA, otherwise known as the reverse passive hemagglutination (RPHA) test, has been used for detection of antigen in patient's serum or body fluids. This chapter describes the in-house design and optimization of the RPHA for *Cryptosporidium* antigen detection in human stool by Farrington et al.

The principle of RPHA is similar to IHA except that instead of antigen, a polyclonal or monoclonal antibody is coated on the erythrocyte. The treated RBCs can then be used to detect the corresponding antigen in test specimen.

Reagents and Equipment Required

Reagents: anti-*Cryptosporidium* oocyst monoclonal antibody in ascitic fluid; mouse kappa IgM; sheep RBC; sheep anti-mouse globulin; normal mouse serum; boric acid; normal saline; PBS; PBS-azide complement fixation test buffer; piperazine buffer; α–chymotrypsin; chromic chloride; sodium hydroxide; sucrose; bovine serum albumin; glutaraldehyde.

Equipment and labware: centrifuge; millipore filter (0.22 μm); dialyzer; freeze-dryer; water bath; 96-well microtiter plate; plate shaker.

The Development of RPHA Test

Antibody Preparation

Anti-*Cryptosporidium* oocyst monoclonal antibody (mAb; MAB-Cl; batch 11) in ascitic fluid, an IgM antibody of the kappa isotype, is obtained from the Central Public Health Laboratory (Colindale, UK).

- Ascitic fluid containing IgM is centrifuged and filtered through a 0.22-μm millipore filter.
- IgM is precipitated using 20 volumes of 2% boric acid solution at 4°C.
- The preparation is centrifuged at 9,000 rpm for 15 minutes.
- The resultant precipitate is suspended in normal saline and dialyzed overnight against saline.
- Antibody-RBC control is prepared by coupling sheep RBCs with mouse kappa IgM (Sigma: ABPC 22; no. M-7394), while uncoupled erythrocytes are used as RBC control.

Antibody Coupling with RBC and Stabilization

- The sheep RBCs are washed with PBS, pH7.2.
- The washed RBCs are incubated with α-chymotrypsin in acidified PBS-azide complement fixation test buffer for 20 minutes at 37°C.
- A chromic chloride coupling solution (stock solution) is prepared by making a 1% solution of $CrCl_3{:}6H_2O$ in normal saline (pyrogen free) and adjusting the pH to 5.0 using a 5M NaOH solution.
- Coupling is carried out using 0.5mg/mL of purified antibody and sheep RBCs in piperazine buffer and adding 1:180 dilution of chromic chloride solution.
- Confirmation of the coupling is done by titrating the RBCs with sheep anti-mouse globulin.
- Stabilization of the coupled RBCs is done by adding 1 mL of the RBC preparation to 100 μL of glutaraldehyde solution (30 μL in 5 mL of PBS) and then freeze-drying overnight in PBS containing 1.5% sucrose and 1.0% bovine serum albumin.
- The cells are reconstituted before use by mixing with an equal volume of PBS and keeping at 4°C for 24 hours.

Test Procedure

Specimen: A stool sample to be tested.

Processing of stool specimen:

- The fecal sample is suspended in PBS at 25% w/v ratio.
- It is heated for 3 minutes in a boiling water bath and then allowed to cool and settle at room temperature for 15 minutes.
- The supernatant is taken and diluted 1:20 with PBS containing 0.1% normal mouse serum.

Performing the test:

- A serial dilution of the processed stool specimen is carried out using a 96-well microtiter plate starting from 1:20 to 1:1,280 using PBS containing 1% normal mouse serum.
- Equal amounts of reconstituted RBCs are added to the wells, and the mixture is shaken and then left undisturbed for 30 minutes before reading the results.

Results and interpretation: A positive result is indicated by a diffuse carpet of agglutinated cells at the bottom of the well, while in a negative test there will be a compact button of cells (Figure 7.2). According to Farrington et al., a test was valid if the titer in the dilution series that began at 1:20 was two or more dilutions higher than the highest titer in either of the control dilution series starting with 1:20.

Quality assurance: With each test run, antibody-erythrocyte control and the uncoupled erythrocyte control should be run in parallel with each dilution. Known positive and negative controls should also be included while testing.

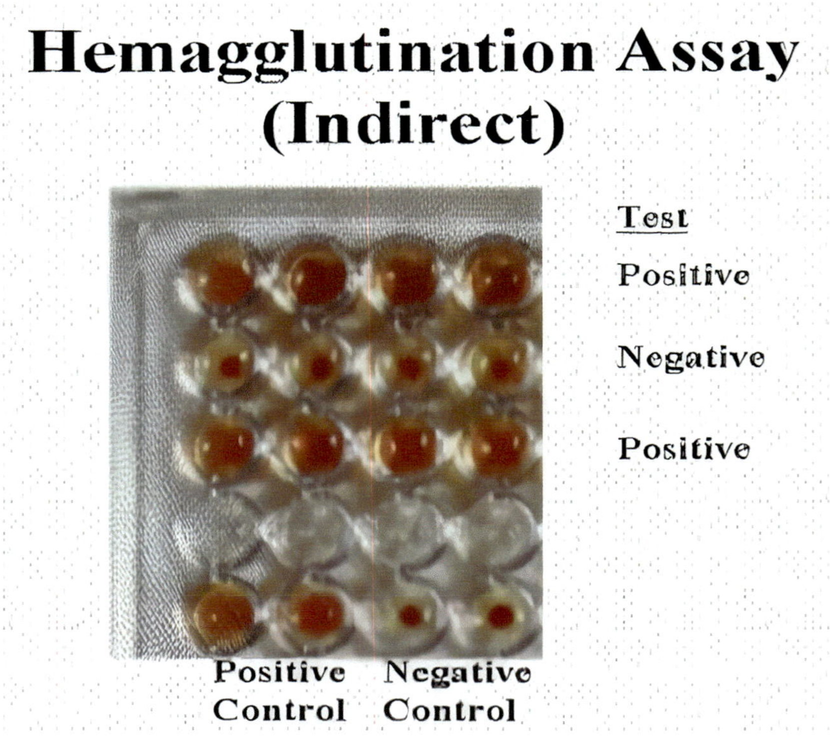

FIGURE 7.2 Indirect hemagglutination assay showing positive and negative results.

(Image courtesy of "Diagnosis of Chagas Disease," https://kdna.net/parasite_course-old/cruzi_files/subchapters/diagnosis.htm.)

Design and Optimization of the Latex Agglutination Test for *Toxoplasma* Serum Antibody Detection

Latex particles are uniformly sized polystyrene microspheres ranging in size from 0.05 μm to 0.48 μm. The term *latex* is used because the process of preparation of these particles evolved from synthetic rubber production, and the emulsion has a milky appearance.

Several kinds of latex particles with specific functional groups are available that can bind with protein molecules via covalent bonding. Many of these have been used to develop immuno-diagnostic tests. Carboxylated latex particles allow covalent binding and have been extensively used, but the COOH groups first require chemical activation. On the other hand, those with aldehyde groups can bind directly with the proteins. Plain polystyrene latex particles have no or ultra-low carboxyl groups on its surface and are suitable for physical adsorption of antibodies or antigens. By contrast, particles with functional groups like carboxyl, amino, chloromethyl, aldehydic, epoxy, or hydroxylate are suitable for covalent coupling.

The simplest method for attaching proteins to particles is passive adsorption.However, proteins may gradually desorb from the particles over time.This can alter the agglutination characteristics of sensitized latex. Additionally, some molecules do not adsorb onto latex particles. Covalent bonding can overcome these problems.

The principle of the latex agglutination test (LAT) is similar to IHA except that latex particles are used instead of erythrocytes.

Reagents and Equipment Required

Reagents: latex particles (1.10-μm diameter; Eagle minimum essential medium (MEM); fetal bovine serum; sodium heparin; PBS (pH 7.6 and 7.4) or MES (2-[n-morpholino]-ethane-sulfonic acid); EDAC (1-ethyl-3-[3 dimethylamino-propyl] carbodiimide hydrochloride).

Equipment and labware: tissue culture system (CO_2 incubator, tissue culture flasks, inverted microscope); counting chamber (hemocytometer); polycarbonate filter (3.0μm); polypropylene tubes; ultrasonicator; centrifuge rotary shaker; automated micropipettes.

Design and Optimization of the Latex Agglutination Test

The following procedure for development of a LAT for toxoplasmosis is based on the work by Mazumder et al.

Preparation of Antigen

Toxoplasma gondii RH strain tachyzoites was used for antigen preparation. They were maintained by twice-weekly passage in human foreskin fibroblast culture.

- The cultures are infected multiple times within 72 hours at 37°C in Eagle MEM containing 3% fetal bovine serum.
- The culture medium is replaced after 24 hours with fresh MEM with 3% fetal bovine serum.
- The culture medium, containing 1 x 10^7 to 2 x 10^7 organisms/mL, is collected. The tachyzoites are filtered through a 3.0-μm polycarbonate filter into a polypropylene tube containing 50 μL of sodium heparin (5,000 U/mL).
- The tachyzoites are then washed thrice in 0.1 M PBS (pH7.4) containing 50 U of sodium heparin/mL and re-suspended in PBS to give a final concentration of 5 x 10^8 organisms/mL.
- Ultrasonication at 125 W/cm^2 for 10 minutes at 4°C is done, and the material is centrifuged at 11,000 rpm for 5 minutes.
- The supernatant containing 1.6 mg of total protein/mL is used as antigen.

Latex Particle Sensitization with the Antigen

- 1.1-μm carboxylated latex particles are used for the test of covalent attachment of the antigen.

- A 1% (v/v) latex particle is activated for 18 hours at 4°C in 0.1 M PBS (pH 7.6) containing 2 mg of EDAC (1-ethyl-3-[3 dimethylaminopropyl] carbodiimide hydrochloride)/mL.
- The adsorption reaction is stopped by centrifuging the suspension at 7,000 rpm for 15 minutes, and the pellet is added to 1% (vol/vol) PBS containing 50µL/mL of *Toxoplasma* antigen.
- The mixture is allowed to react for 2 hours at room temperature in a shaker at 40 to 50 oscillations per min.
- Bovine serum albumin is then added to a 1% final concentration to block unreacted binding sites on the latex.
- The toxoplasma-sensitized latex reagent can be stored at 2°C to 8°C.

  Note: While designing a test, it should be kept in mind that PBS under acidic conditions typically gives poor result. MES (2-[n-morpholino]-ethanesulfonic acid) buffer is a better alternative in such situations.

Test Procedure

Specimen: Serum samples are to be tested.

Performing the Test:

- A 1:4 dilution of the patient's serum in 0.1 M PBS (pH 7.4) containing 5% bovine serum albumin is used for the test to minimize nonspecific agglutination.
- Equal volumes (25 µL) of the serum sample and the sensitized latex particles are mixed on a glass slide by a plastic applicator, and manual rotation of the slide is done by hand for 3–5 min.
- Clumping of the particles is determined visually under a high-intensity lamp.

  Note: The authors noted that the specificity of the test was 100% and sensitivity was 94% and 86% when compared with commercial indirect fluorescence test and ELISA, respectively.

Results and interpretation: Agglutination within 2 minutes of addition of latex reagent and absence of agglutination in the control half indicate a positive test. Reactions that seem to be "curdy" or "stringy" may not indicate a true-positive result (Figure 7.3).

Quality assurance: Positive and negative controls were used during each test performed. If autoagglutination in the control half is observed, the test is invalid and must be repeated.

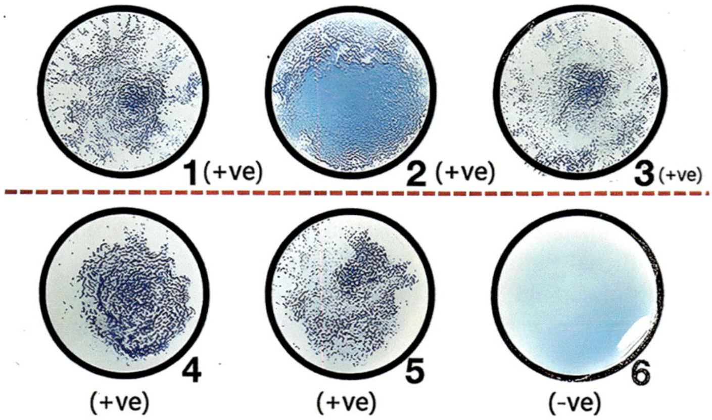

FIGURE 7.3 Results and interpretation of latex agglutination test. Some reactions are reported as positive or negative, and other reactions are graded on a 1+ to 4+ scale, with 2+ usually the minimum amount of agglutination visible in a positive sample without the aid of a microscope.

(Photo courtesy of "Microbe Notes," https://microbenotes.com/latex-agglutination-test/.)

Designing and Optimizing the Co-agglutination (Co-A) Test for *Echinococcus* Antigen Detection in Urine and Serum

Staphylococcus aureus Cowan 1 strains have a Protein A on its surface, which can naturally adsorb the F_C portion of antibody molecules against *Echinococcus* antigen, leaving the antigen-binding F_{ab} sites open. When these echinococcal antibodies–coated organisms are reacted with the patient's sample containing the corresponding echinococcal antigen, clumping of the organisms occurs.

The following procedure outlines the development of the co-agglutination (COA) test for antigen detection in hydatid disease as adapted from the report by Parija and associates. This procedure can be used for detection of any other parasitic antigen in serum, urine, and other medium, replacing the analytical antibody used in this procedure.

Reagents and Equipment Required

Reagents: hydatid fluid antigen; Freund's complete adjuvant; saturated ammonium sulphate (Ph 7.0); saline (Ph 7.0); PBS (pH 7.2); Mueller Hinton agar; sodium azide; formaldehyde solution (1.5%); sodium acetate; ethanol; bovine serum albumin.

Equipment and labware: centrifuge; stirrer; dialyzer; petri dishes; automated micropipettes; syringe and needle.

Others: adult male, white New Zealand rabbit; *Staphylococcus aureus*, Cowan's 1 strain (NCTC 8530 or ATCC 12598).

Design and Optimization of the Co-agglutination Test

Antigen for Raising Hyperimmune Serum

The antigen used is hydatid cyst fluid obtained from humans or animals as described above for antibody detection.

Preparation of Hyperimmune Serum

Two healthy, adult male, white New Zealand rabbits (3–4 kg and 5–6 months age) are used.

- 1.5 mL of hydatid fluid antigen (protein concentration of 8mg/mL) is mixed with an equal volume of Freund's complete adjuvant and injected intramuscularly at a dose of 0.5 mL in all four limbs simultaneously.
- After 6 weeks, rabbits are reinjected with the same dose of Freund's incomplete antigen.
- After 10 days, blood samples are collected by marginal ear vein bleeding. Maximum 3 mL of blood can be collected by this method at a time. If the central ear artery is used for collection, maximum 15 mL of blood can be collected weekly from a 4-kg rabbit.
- The serum is separated and purified as follows: equal volumes (1 mL) of cold serum and cold saline (pH7.0) are mixed, and the mixture is added drop by drop to 2 mL of cold saturated ammonium sulphate (pH 7.0) with constant stirring over a 30-minute period on ice.
- It is then centrifuged at 3,000 rpm at 0°C.
- The precipitate is re-suspended in 2 mL of saline, and the above steps are repeated until the supernatant becomes colorless.
- The final precipitate is suspended in 1 mL and dialyzed against PBS (pH 7.2) so that all ammonium sulphate is removed.

Staphylococcus aureus *Cells*

Cowan's 1 strain of *S. aureus* bearing protein A is used (NCTC 8530, ATCC 12598).

- Cells are grown on Mueller-Hinton agar at 37°C for 18 hours and then harvested.
- The cells are suspended in PBS and centrifuged at 4,900 rpm for 10 minutes.

- The pelleted cells are then washed three times in PBS (pH 7.2) containing 0.05% sodium azide.
- To fix the cells, the pellet is suspended in 10 vol of PBS with 1.5% formaldehyde at room temperature for 90 minutes.
- The cells are then again re-washed thrice in PBS containing sodium azide and then heated at 80°C for 5 minutes in a water bath.
- Two washes with PBS are then made and a 10% suspension of cells in PBS with sodium azide is prepared.

Sensitization of the Cells

The cells should be sensitized immediately after preparation.

- One milliliter of the 10% suspension of cells is added to 0.1 mL of hyperimmune antiserum, mixed well, and kept at room temperature for 30 minutes.
- This is followed by further washing with PBS, after which the cell pellet is suspended in PBS and a 2% suspension of cells is made in PBS with 0.1% sodium azide.
- This can be stored at 4°C. A 2% suspension of unsensitized cells is used as control.

Test Procedure

Specimen: Urine sample to be tested.

Processing of urine specimen

Urine concentration: Before using the urine sample for testing, it is preferable to concentrate it, since the amount of antigen is low.

- Ethanol concentration technique may be used for this purpose.
- In this method, 1 mL of urine is mixed with 0.1 mL of a cold solution made up of 3 M sodium acetate, 0.1% bovine serum albumin, and 0.1% (wt./vol) sodium azide.
- To this, 1.5 mL of 96% ethyl alcohol at −20°C is added.
- The mixture is vortexed and then cold centrifuged at 9,000 rpm for 10 minutes at 4°C.
- After discarding the supernatant, the deposit is resuspended in 0.1 mL of PBS (pH 7.2).

The test is performed as follows:

- A clean glass slide is marked out into two halves with a glass-marking pencil/pen.
- One drop of urine or serum is added to each half.
- One drop of 2% sensitized cells is added to one half of the slide and one drop of unsensitized cells to the other half, followed by manual rotation.

Results and interpretation: In a positive test, there will be large visible clumping of the cells with clearing of the suspension within 2 minutes. Unsensitized cells will show no clumping.

Quality assurance: A known positive and a known negative urine or serum sample should be run in parallel.

Limitations of Particle-Based Agglutination Assays

Particle agglutination tests are easy to perform, economical, can be carried out at room temperature, and require no wash step. However, they have the following limitations:

- Subjective interpretation, which can lead to inter-observer variability, particularly for weak agglutination patterns if titers need to be determined.
- Immunoglobulin classes cannot be determined.
- If pure antigen fraction is not used, it frequently produces false-positive results. This is particularly true for helminth parasites or other parasites that cannot be cultivated.

- Prozone/post-zone phenomenon can occur if there is an antigen/antibody excess in the patient's sample. These excess antibodies/antigens bind to the antigenic/antibody sites in a manner leading to inhibition of the lattice formation, with no agglutination. This false-negative reaction can be prevented by testing with diluted serum to bring down the concentration of the analyte.

CONCLUSION

Carrier particle–based tests are one of the oldest immunological test still used today. Latex- and hemagglutination tests are popular tests that are quick and easy to perform. Specificity of these tests can be quite high if an appropriate antigen/antibody is used, though their sensitivity is variable. One of the problems with these tests is the subjective interpretation of results by the observer. Several experimental particle agglutination tests have been developed, but only a few are available commercially for parasitic diseases.

FURTHER READINGS

Farrington M, Winters S, Walker C, Miller R, Rubenstein D. Cryptosporidium antigen detection in human feces by reverse passive hemagglutination assay. *J Clin Microbiol.* 1994;32:2755–2759.

Farshy DC, Kagan IG. Use of stable sensitized cells in indirect microhemagglutination test for malaria. *Am J Trop Med Hyg.* 1972;21:868–872.

Karen DF, Warren JS. *Diagnostic Immunology.* Baltimore, MD: Williams and Wilkins; 1992b.

Katti MK. Ideal carrier particles for agglutination tests. *J Clin Microbiol.* 1999;37:2120.

Mazumder P, Chuang HY, Wentz MW, Wiedbrauk DL. Latex agglutination test for detection of antibodies to *Toxoplasma gondii. J Clin Microbiol.* 1988;26:2444–2446.

Ortega-Vinuesa JL, Bastos-González D. A review of factors affecting the performances of latex agglutination tests. *J Biomater Sci.* 2001;12:379–408.

Pagnozzi D, Biosa G, Addis MF, Mastrandrea S, Masala G, et al. An easy and efficient method for native and immunoreactive *Echinococcus granulosus* antigen 5 enrichment from hydatid cyst fluid. *PloS One.* 2014;9(8):e104962.

Parija SC, Ananthakrishnan N. Evaluation of stabilized cells in the indirect haemagglutination test for echinococcosis. *J Med Microbiol.* 1985a;19:95–98.

Ravinder PT, Parija SC, Subba Rao KSVK. Urinary hydatid antigen detection by coagglutination, a costeffective and rapid test for diagnosis of cystic echinococcosis in a rural or field setting. *J Clin Microbiol.* 2000a;38:2972–2974.

Sarkari B, Rezaei Z. Immunodiagnosis of human hydatid disease: where do we stand? *World J Methodol.* 2015;5:185–195.

Turgeon ML. *Immunology and Serology in Laboratory Medicine.* 2nd ed. St Louis, MO: Mosby; 1996.

8 Precipitation Reaction-Based Tests

INTRODUCTION

Precipitation reaction can be defined as a reaction between a soluble antigen and its antibody in the presence of electrolytes at a suitable temperature and pH resulting in an antigen–antibody complex formation as an insoluble precipitate. These tests do not use any particles except soluble antigens that precipitate in the presence of corresponding antibodies in a patient's serum. This contrasts with particle-based agglutination tests, described in the previous chapter, that use insoluble carrier particles coated with antigen. When a specific antigen–antibody reaction occurs, the particles clump together, producing visible agglutination.

A significant number of precipitation tests have been described in the diagnosis of infectious diseases, both with and without the application of an electric field. However, merely two such tests have been developed and evaluated specifically for detection of parasitic diseases.

PRINCIPLE OF PRECIPITATION REACTION-BASED TESTS

Antigens and antibodies have positive and negative polar groups on their surfaces in specific but reciprocal patterns. When the corresponding antigen and antibody molecules are mixed, electrical attraction and repulsion take place. This results in a specific orientation of these molecules according to their molecular shapes and charges forming a closely fit model. Precipitation occurs rapidly when antigen and antibody are present in optimum proportion. If the amount of precipitate in different tubes is plotted, the resulting curve will have three different phases:

1. A zone of antibody excess (*prozone*) in which some uncombined antibody will be present.
2. A zone of equivalence in which all antigen and antibody are completely precipitated.
3. A zone of antigen excess (*post zone*) in which all antibodies have combined with antigen but some uncombined antigen will be present.

TYPES OF PRECIPITATION REACTIONS

A precipitation reaction can take place in a liquid or semisolid medium or in gels such as agar gels, agarose, or polyacrylamide gels. When the reaction occurs in a liquid medium, formation of a ring, or flocculation, takes place, while in a gel format, a line of precipitation (precipitin line) is formed. Sometimes electrical current is applied to expedite the movement of antigen or antibody in a gel, resulting in rapid precipitation.

The precipitation tests are of the following types:

Precipitation in a liquid medium

1. Ring test: A ring of precipitate forms at the junction of two fluids of antigen and antibody.
2. Slide flocculation test: Formation of floccules of antigen and antibody on a slide.
3. Tube flocculation test: Formation of floccules of antigen and antibody in a tube.

DOI: 10.1201/9781003644224-11

Gel precipitation tests

These are considered more sensitive and specific than precipitation in a liquid medium.

1. Single diffusion in one dimension (Oudin procedure).
2. Double diffusion in one dimension (Oakley Fulthorpe procedure).
3. Single diffusion in two dimensions (radial immunodiffusion or Mancini method).
4. Double diffusion in two dimensions (Ouchterlony procedure).

Gel precipitation in an electrical field

1. Immunoelectrophoresis.
2. Countercurrent immunoelectrophoresis (CIEP).
3. One-dimensional single electro immunodiffusion (Rocket electrophoresis).
4. Two-dimensional immunoelectrophoresis (Laurell's procedure).

PRECIPITATION REACTION-BASED TEST FORMATS FOR ANTIGEN AND ANTIBODY DETECTION

Most of these formats have been used for detection of antigen or antibody in different situations and have their own merits and demerits. Two techniques that have been employed frequently for diagnosis of parasitic infections in experimental settings are the Ouchterlony procedure and countercurrent immunoelectrophoresis (CIEP).

Double Diffusion in Two Dimensions (Ouchterlony Procedure)

Örjan Ouchterlony, a Swedish physician, developed the test in 1948. Here, agarose gel is poured on a slide or petri dish and holes are punched, one at the center and others along the periphery in a ring-like fashion. For antibody detection, the central well is filled with the antigen solution and the peripheral wells with the patient's sera along with known negative and positive control sera. It is then incubated and the lines of precipitation are noted. Two types of patterns may be obtained:

a. Two precipitin lines totally combine. This pattern is called *reaction of identity* and indicates that the antibodies in the adjacent wells are identical (positive control and test serum).
b. Two precipitin lines fuse but form a spur-like projection. This is known as *reaction of partial identity* and signifies that the test serum is showing cross-reactivity.
   Note: A third *pattern of non-identity* is found when a known antibody is kept at the central well surrounded by different antigens in the surrounding wells.

Countercurrent Immunoelectrophoresis

This can be considered a modification of the Ouchterlony procedure, except that the test can be completed much faster (within 30–45 minutes) by application of an electric current. It depends on the movement of the antibody toward the cathode and the antigen towards the anode through the agar in the electric field. At the area where optimal concentration is reached, they form the line of precipitation. The antibody fraction should be kept at the anodic end while the antigen is at the cathodic end. Agarose gel with high electro-endosomatic flow should be used for the test.

APPLICATIONS IN PARASITIC DIAGNOSIS

Immunodiffusion and countercurrent immunoelectrophoresis (CIEP) are antigen–antibody precipitation techniques used in parasitology, mainly for serodiagnosis. Immunodiffusion is simple, inexpensive, and highly specific and is useful for confirmation and retrospective diagnosis.

It is, however, less sensitive compared with newer immunoassays. Immunodiffusion has various applications in the diagnosis of parasitic infections. It has been used for detection of serum antibodies in hydatid disease (detection of hydatid antibodies, with arc-5 band diagnostic), visceral leishmaniasis, amoebiasis, filariasis, schistosomiasis, and trichinosis. It is also used for antigen detection in hydatid disease, visceral leishmaniasis, and invasive amoebiasis such as amoebic liver abscess.

CIEP provides rapid results (within 30–60 minutes) and is more sensitive than the simple immune diffusion. The test is useful when quick confirmation is needed, as well as and in early and active infections, especially when parasites are not easily demonstrable.

CIEP has been used for circulating serum antigen in hydatid disease, filariasis, visceral leishmaniasis, and amoebic liver abscess. The test has also been used for urine antigen detection excreted in visceral leishmaniasis and filariasis, as well as for stool antigen detection for the intestinal amoebiasis and giardiasis.

The in-house development and design of representative antibody and antigen detection methods using immunodiffusion and CIEP include primarily:

1. Designing the double diffusion for the detection of serum antibody in onchocerciasis.
2. Designing the CIEP for detection of serum antigen in amoebic liver abscess.

Designing the Double Diffusion Test for Serum Antibody in Onchocerciasis

Immunodiffusion is a serological technique based on the passive diffusion of antigen and antibody in agar gel, resulting in the formation of a visible precipitin line at the zone of equivalence. This chapter describes the in-house design and optimization of the double diffusion test protocol for detection of serum antibody in onchocerciasis described by Takaoka et al.

Reagents and Equipment Required

Reagents: acetone; normal saline; PBS (pH: 7.2); Veronal buffer (pH: 8.6); Freund complete adjuvant; agarose; Coomassie brilliant blue; double-distilled water.

Equipment and labware: tissue homogenizer; centrifuge; magnetic stirrer; glass slides; micropipettes; agar punching device; needle and syringe.

Others: adult worm of *Onchocerca volvulus*; adult rabbits.

Development of the Immunodiffusion Test

The development of the immunodiffusion test occurs in the following steps:

Antigen Preparation

Adult *O. volvulus* worms are teased out of the nodules removed from the patient.

- The isolated worms are washed three times with normal saline and then homogenized with cold acetone to remove the lipids.
- The preparation is then centrifuged at 3,000 rpm for 30 minutes.
- The supernatant is discarded and the sediment is added to 0.15 M PBS (pH 7.2).
- The suspension is homogenized further for 10 minutes at 0°C and then stirred slowly with a magnetic stirrer for 48 hours at 4°C.
- The suspension is centrifuged at 8,000 rpm for 30 minutes at 4°C.
- The supernatant is collected and concentrated using a collodion bag, and the resultant antigen is stored at −20°C until used.
- Protein concentration of antigen is standardized at 3mg/mL.

Immunized Rabbit Serum

The rabbit serum is prepared to get positive control.

- The prepared antigen as mentioned above is mixed with an equal volume of Freund complete adjuvant.
- The mixture is inoculated by subcutaneous route into foot pads of two rabbits, twice a week for 3 consecutive weeks.
- One week after the last injection, blood is collected from the rabbits and serum is separated.

Preparation of Plates

- A 1% agarose mixture is prepared in 0.05 M veronal buffer (pH8.6) with ionic strength of 0.05.
- Four milliliters of agarose mixture is layered carefully on a clean glass slide. Once the agar solidifies, seven wells, each 3 mm in diameter, are made 5 mm apart in the agar with an agar punch, along with a central well.

Test Procedure

Specimens: Serum sample to be tested.

- The antigen is kept in a central well.
- The patient's serum to be tested and rabbit hyperimmune serum are kept in adjacent wells surrounding the antigen well.
- The slide is kept in a wet chamber for 3 days at 4°C with saline wash on each day.
- The plates are treated with Coomassie brilliant blue to stain the precipitin bands.

Results and interpretation: In a patient positive for onchocerciasis, lines of identity could be found between the unknown sample and the positive rabbit serum sample (Figures 8.1 and 8.2).

Quality assurance: The immunized rabbit serum is used as positive control, while serum from a healthy person can be included as negative control. The agarose should be of superior quality to allow easy diffusion of molecules, and the slide should be kept in a humid environment to prevent gel drying.

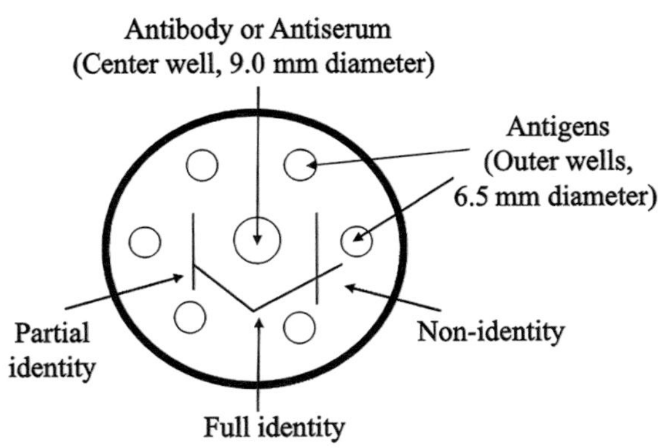

FIGURE 8.1 Ouchterloney double diffusion test.

(Photo credit: Mujtaba MG, Baliban T, Bhagu J, Herrera M. A laboratory exercise simulating antibody and antigen reactions of the Ouchterlony double immunodiffusion assay using inorganic salts. *J Microbiol Biol Educ*. 2021;22(2): e00103-21; distributed under the terms of the Creative Commons Attribution- 4.0 International license.)

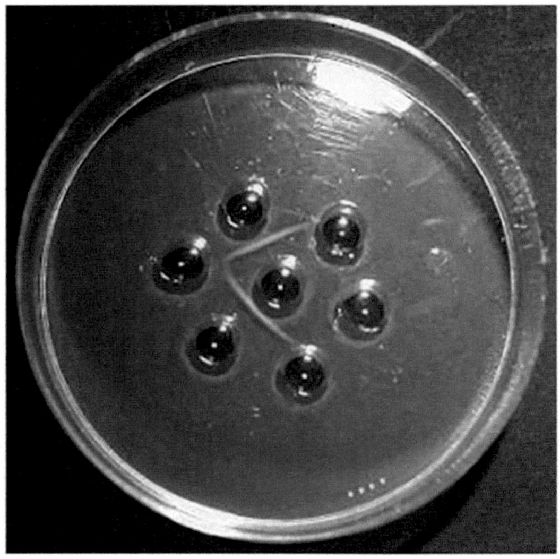

FIGURE 8.2 Ouchterloney double diffusion test showing lines of identity.

(Photo credit: https://bioresource.in.)

Designing and Optimizing the Countercurrent Immunoelectrophoresis for Detection of Serum Antigen in Amoebic Liver Abscess

This section describes the in-house design and optimization of the CIEP protocol for detection of serum antigen outlined by Parija and associates.

Reagents and Equipment Required

Reagents: dimethyl barbituric acid; sodium dimethyl barbiturate; sodium azide; agarose or Bacto agar (Difco); Whatman 3 filter paper; glacial acetic acid; methylated spirit; Amido black; glycerol; double-distilled water; Freund complete adjuvant.

Equipment and labware: water bath; clean glass slide or glass plate; electrophoresis chamber with power pack; agar punching device; incubator; micropipettes.

Others: adult rabbits (weighing 3–4 kg each); *Entamoeba histolytica* (NIH:200 strain).

Design and Optimization of the CIEP Test

Preparation of Hyperimmune Amoebic Antibody Sera

Antigen

- *Entamoeba histolytica* (NIH:200 strain) is grown axenically.
- Amebae from a 48-hour culture are removed and washed thrice in sterile normal saline.
- The organisms are concentrated by centrifugation at 2,000 rpm for 30 minutes.
- The concentrated product is sonicated in the cold for 5 minutes in an ultrasonic disintegrator with 1-minute bursts and 2-minute breaks between the bursts.
- The sonicated product is centrifuged at 9,000 rpm for 30 minutes, and the supernatant is collected and used as the antigen.

Hyperimmune Serum

- The antigen prepared as above is mixed with an equal volume of Freund complete antigen.

- Half a milliliter of the mixture is injected intramuscularly in each limb of an adult rabbit.
- A repeat dose is given after 6 weeks.
- After another 10 days, blood is collected and the serum separated.
- The serum is tested for presence of anti-amoebic antibody by indirect hemagglutination test (the titer in the study was found to be 1,024).

Veronal Buffer Preparation

Dimethyl barbituric acid, 276 g is dissolved in 1 L of distilled water by gentle heating.

Sodium dimethyl barbiturate (15.45 g) and sodium azide (1 g) are added to it. This generates 0.075 M Veronal buffer (pH 8.6).

Agarose Preparation

0.1% agarose or 1% Bacto agar (Difco) is prepared by adding 1 g of agarose/agar to 10 mL of veronal buffer and 90 mL distilled water and melted in a boiling water bath.

Test Procedure

Specimens: Serum sample to be tested.

- Agarose/agar, 5 mL is pipetted onto a clean glass slide (50 mm x 75 mm) and allowed to solidify for 15 minutes. It is then kept at 4°C overnight in a refrigerator.
- Using an agar punch, six pairs of wells, 4 mm in diameter and 3 mm apart, are cut according to the template toward the two ends of the slide.
- The wells are filled with 10 µL of hyperimmune serum at one end and 10µL of serum to be tested (containing the suspected antigen), along with positive and negative control serum, at the opposite end of the slide.
- An electrophoresis tank is filled with the veronal buffer, and the slide is carefully kept horizontally in the chamber so that the buffer just covers the gel surface. Alternatively, Whatman 3 filter paper strips are connected to the buffer solution and agarose.
- Wells containing the hyperimmune sera are kept at the anodic end of the electrophoresis chamber while the test sample wells at the cathodic end.
- 50 V or 10–15mA electricity is applied, and the electrophoresis is allowed to continue for about 30 minutes.
- The slide is removed, and lines of precipitation are observed in oblique light against a black background.
- Unprecipitated protein is washed out of the agar by immersion of the slide for in veronal buffer for 24 hours.
- Excess salt is removed by washing with 1% acetic acid for 15 minutes.
- Staining is done for 30 min with 1% amido black solution in a solvent made of 1 mL glacial acetic acid, 40 mL distilled water, and 50 mL methylated spirit.
- Excess stain is the washed with the solvent to give a final preparation with dark bluish precipitation bands on a clear background.
- The slide is soaked in 1% acetic acid containing 1% glycerol for 15 minutes and dried at 37°C in an incubator.

Results and interpretation: A precipitin line will be visible between the antigen and antibody wells after 30 minutes in a positive case (Figure 8.3). The precipitin line should also be visible in the control row where known antigen and corresponding antibody have been kept in the wells. In a negative case where no antigen is present, there will be no precipitin line.

Quality assurance: Known antigen and its corresponding antibody should be run on the same plate/slide. Antigen should always be kept at the cathodic end while antibody at the anodic end.

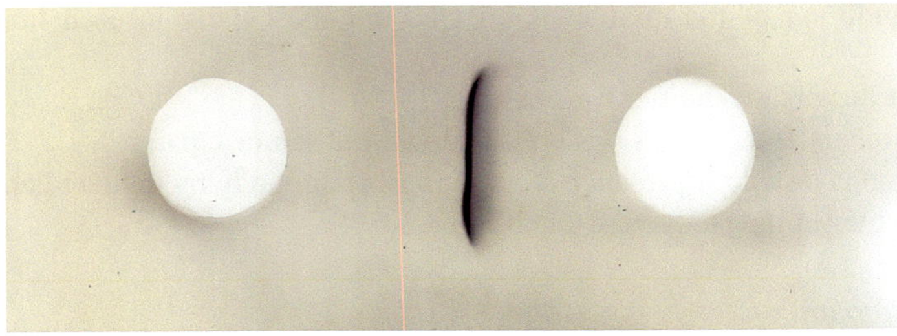

FIGURE 8.3 Countercurrent immunoelectrophoresis showing the line of precipitation.

(Source: Wikimedia Commons. Author: Ernst Hempelmann. This file is made available under the Creative Commons CC0 1.0 Universal Public Domain Dedication.)

CONCLUSION

Both double diffusion in two dimensions and CIEP are qualitative methods that can detect the presence or absence of the antibody or antigen in patient's serum or other bodily fluids. As such, they can be used as a screening test. It has been observed that the CIEP is 10–20 times more sensitive than the Ouchterlony immunodiffusion technique. Immunodiffusion is mainly used as a confirmatory serological test in parasitic infections, particularly in hydatid disease, and has largely been replaced by more sensitive techniques. The ease of both test procedures makes them attractive options for developing in-house tests for diagnosis of common or uncommon parasitic infections. However, careful selection of the antigen is imperative to avoid cross-reactions with related or unrelated parasites.

FURTHER READINGS

Parija SC, Karki BMS. Detection of circulating antigen in amoebic liver abscess by counter-current immunoelectrophoresis. *J Med Microbiol.* 1999;48:99–101.

Shariff GM, Parija SC. Counter-current immunoelectrophoresis test for serodiagnosis of hydatid disease by detection of circulating hydatid antigen. *J Microbiol Methods.* 1991;14:71–76.

Takaoka M, Lujan TA, Hashiguchi Y, Kawabata M, Ito Y, Hayashi S. Evaluation of the double diffusion test for the serodiagnosis of onchocerciasis in Guatemala. *Jap J Parasit.* 1983;32:451–457.

9 Complement Fixation Test

INTRODUCTION

Complement fixation test (CFT) was one of the first antigen–antibody tests used for diagnostic purposes. It was introduced by Bordet and Gengou in 1901 for serodiagnosis of typhoid fever and in 1906 by Wassermann for syphilis. Ghedini in 1907 first used it for serological diagnosis of hepatic hydatid and investigated *Ancylostoma* and *Ascaris* infestations using this test.

For a long time, CFT was considered a highly specific and sensitive diagnostic test for a variety of bacterial, parasitic, viral, and fungal infections and was often taken as the reference test. Although the test appears to be relatively simple, the technique is complicated and requires rigorous methodology and stringent standardization of antigen amount, complement, and sensitized RBCs. From the 1970s onward, its use has steadily declined, but it is nevertheless a powerful technique and can be used to design in-house diagnostic tests.

PRINCIPLE OF THE COMPLEMENT FIXATION TEST

CFT requires four reagents: an antigen, patient's serum containing the antibodies, fresh complement from animals (commonly guinea pig), and a hemolytic system (sensitized sheep or human red cells).

The test is performed in two stages, as follows.

Complement Fixation Step

The patient's serum is allowed to react with the antigen followed by the addition of the complement. If the patient's serum contains homologous antibody, it will react with the antigen, and the complement in the system will get anchored or "fixed" to the antigen–antibody complex and be removed from the reaction mixture. In the event of dissimilar antigen and antibody, the complement will not get fixed and will remain free in the mixture. This step can be performed at 37°C for 1 hour, at room temperature for several hours, or in an icebox at 0°C overnight, the last procedure being considered the most sensitive. Complement molecules normally deteriorate rapidly at higher temperatures.

Indicator Hemolytic Step

To visualize the actual fixation or attachment of the complement to the antigen–antibody complex, it is necessary to demonstrate that no free complement is any longer present in the reaction system. For this, an indicator or hemolytic system consisting of sensitized sheep or human red cells are used. The sheep RBCs are sensitized by using rabbit anti-RBC antibodies, and if it meets the free complement, the RBCs will lyse. Hence, if the antigen–antibody complex has formed, all the complement molecules get fixed in the previous step. The subsequent addition of the indicator system will show no hemolysis, which is interpreted as a positive test. If homologous antibodies are not present in the patient's serum, the complement will remain free and will cause hemolysis of the RBCs (negative test), as shown in Figures 9.1 and 9.2.

DOI: 10.1201/9781003644224-12

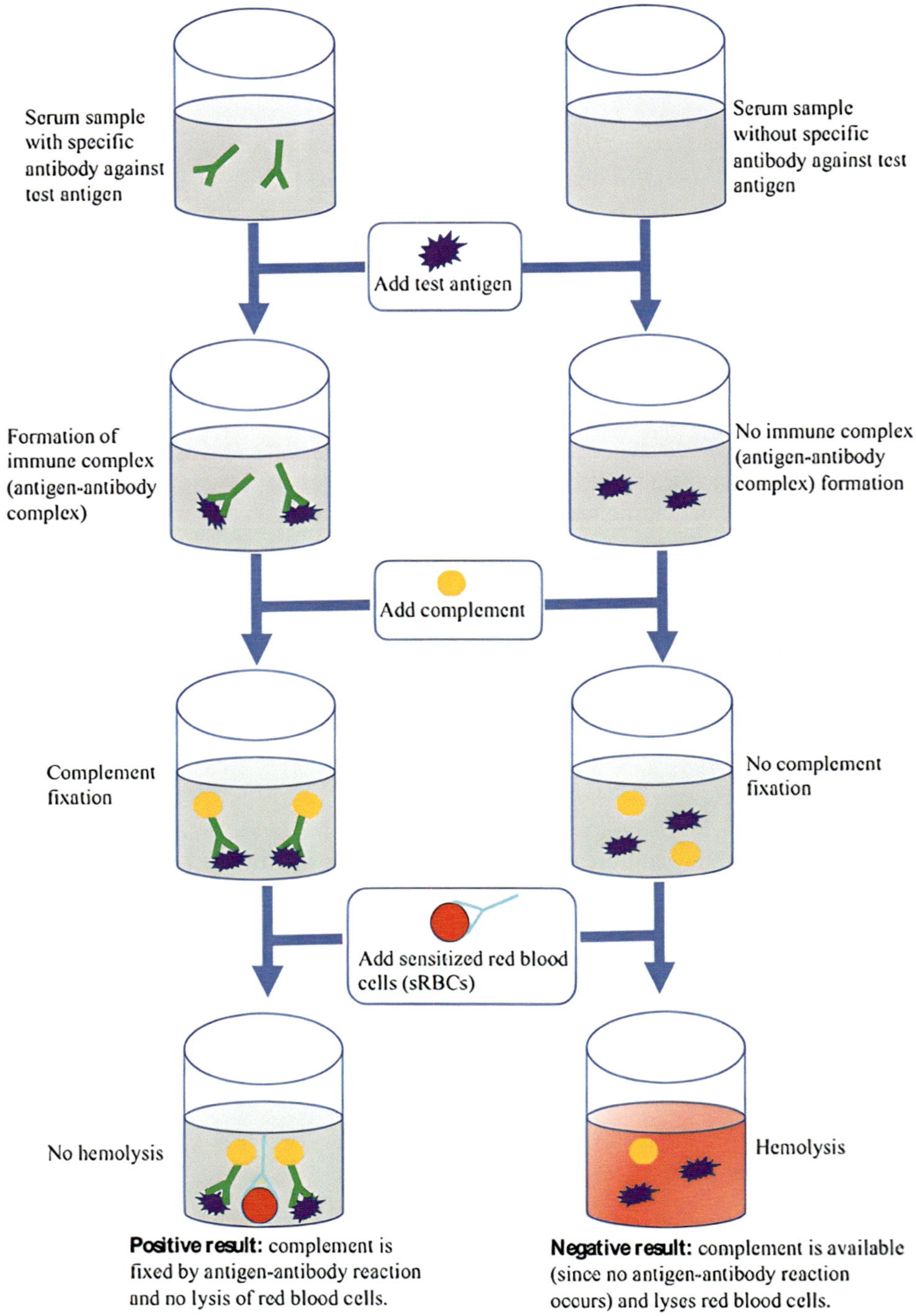

FIGURE 9.1 Principle of complement fixation test.

(Photo credit: Khan LB, Tsai J-YC. A simple, cost-effective undergraduate workshop based on simulated complement fixation test to teach the concept of complement system. *Eur J Health Biol Educat.* 2019;8(1):19–30. Open Access – This article is under the CC BY license (https://creativecommons.org/licenses/by/4.0/.).

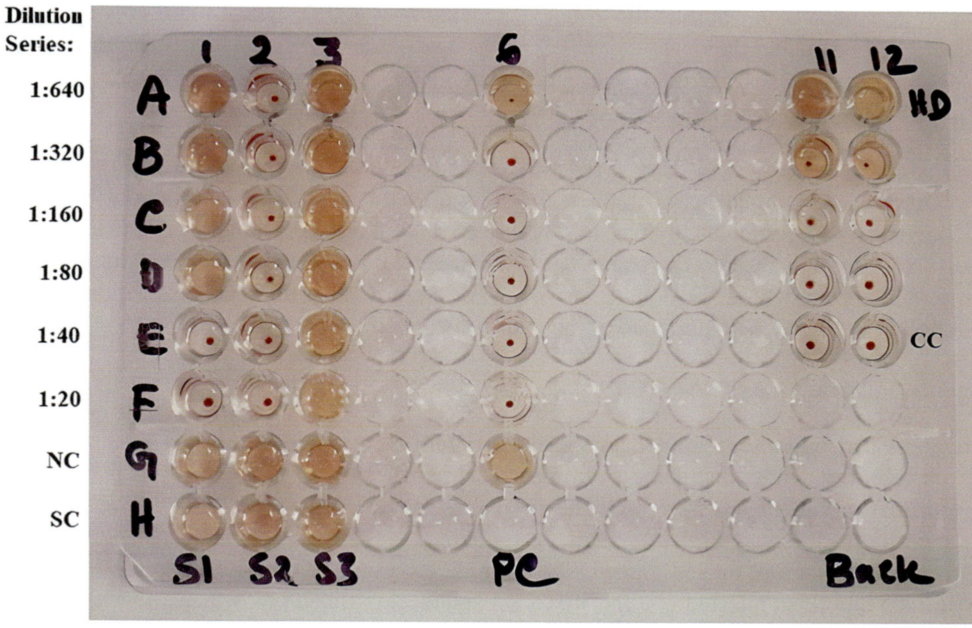

FIGURE 9.2 A microtiter plate–based complement fixation test. Patient's serum samples S1 (titer > 1:40) and S2 (titer > 1:640) showed positive result, and S3 (titer < 1:20) showed negative result for influenza virus A infection. *CC*, cell control (E11–E12); *HD*, hemolysis dose (A11–A12); *NC*, negative control (G1, G2, G3, and G6); *PC*, positive control (A6–F6); *SC*, serum control (H1, H2, and H3).

(Photo credit: Khan LB, Tsai J-YC. A simple, cost-effective undergraduate workshop based on simulated complement fixation test to teach the concept of complement system. *Eur J Health Biol Educat.* 2019;8(1):19–30. Open Access –This article is under the CC BY license (https://creativecommons.org/licenses/by/4.0/.).

COMPLEMENT FIXATION TEST FORMATS FOR ANTIBODY DETECTION

The CFT is a classical serological method used to detect pathogen-specific antibodies in patient serum. The test is based on the ability of antigen–antibody complexes to fix complement; absence of hemolysis in the indicator system shows a positive reaction.

Complement fixation tests have been widely used earlier in the diagnosis of several infectious diseases including viral (influenza and adenovirus), bacterial (*Mycoplasma pneumoniae*, *Chlamydia* species, *Coxiella burnetii*), and fungal (*Histoplasma capsulatum*, *Coccidioides immitis*). They have played an important role in detecting antibodies to these pathogens. Although largely replaced by newer, more sensitive assays, CFTs remain valuable in some reference laboratories and for retrospective diagnosis by demonstrating rising antibody titters.

Applications in Parasitic Diagnosis

The CFT remains of academic importance and illustrates the historical development of serological diagnosis in parasitology. It is a classical serological method used to detect parasite-specific antibodies in patient serum. CFTs have been applied in the diagnosis of parasitic diseases such as pulmonary paragonimiasis, amoebiasis, hydatid disease, toxoplasmosis, and trichinosis. The test may still be used in select laboratories for supportive or retrospective diagnosis.

The following sections outline the in-house development and design of representative antibody detection methods using the CFT.

Complement Fixation Test Protocol

A universal CFT is described here for any parasitic infection, whether protozoan or helminthic, irrespective of the antigen being used, and is based on the report by Bozicevich, Hoyem, and Walston. Here the complement titer is determined based on the amount needed to give 50% hemolysis when compared to the color standard. The test must be performed in a total volume of 1 mL in which four of these 50% units are used. For the test using microtiter plates, the reader can consult other publications mentioned at the end of this chapter.

Reagents and Equipment Required

Reagents: trisodium citrate (4% solution) or sodium citrate (3.8%); Alsever solution; phosphate buffer saline; distilled water; normal saline.

Equipment: Vacutainer or syringe and needle; graduated conical centrifuge tube (15 mL); centrifuge; automated micropipettes; water bath.

Others: sheep blood; pooled guinea pig serum; hemolysin and parasite antigens.

- Sheep blood is collected and the RBCs separated as outlined below.
- Pooled guinea pig serum is prepared as follows:
  - Eight to 10 healthy guinea pigs are taken, consisting of an equal number of males and females, which are free of external parasites.
  - Blood is collected either from lateral saphenous veins or a jugular vein from each animal separately in serum separator tubes or plain blood collection tubes without anticoagulants.
  - The collected blood is allowed to stand undisturbed for 30 minutes in an upright position until a solid clot is formed.
  - It is then centrifuged at 2,000 rpm for 10 minutes in a cold centrifuge (4°C).
  - The supernatant serum from each tube is collected and transferred to individual tubes. It should be clear, straw colored, without any hemolysis.
  - Equal volumes of serum are taken from individual tubes and combined in a larger container to create the pooled serum.
  - The pooled serum is aliquoted into smaller vials.
  - If the serum is to be used immediately, the vials can be stored at 2°C–8°C up to 1 week.
  - For long-term storage, they are kept at −20°C or −80°C.

    Note: Various parasite antigens and all the other biologicals like sheep RBC, hemolysin, and pooled guinea pig serum and complements are available commercially (*vide infra*).

Design and Optimization of the Complement Fixation Test

Color Standard Preparation

To determine the end point of 50% hemolysis, a color standard is prepared using 2% sheep red cells.

- Sheep blood is collected under aseptic conditions by venipuncture of the jugular vein using a Vacutainer or needle (18- or 20-gauge) and syringe. No more than 1% of the animal's blood volume in one collection or over a 24-hour period should be done.
- The blood is collected in citrate anticoagulant or defibrinated after collection.
- For long-term preservation, RBCs should be washed four times in normal saline or preserved in Alsever solution to reduce the fragility and prevent hemolysis during storage.

- A 15-mL graduated, conical centrifuge tube is filled with washed cells and centrifuged at 2,000 rpm for 10 minutes.
- The quantity of packed cells is measured and the volume occupied by the cells is multiplied by a factor of 50 to make a 2% cell suspension. This final volume is made by adding the required amount of saline. Thus if 2 mL of packed cell is obtained, then 98 mL of saline is to be added to give a final volume of 100 mL.
- Ten milliliters of the cell suspension is transferred to a graduated centrifuge tube, and packed cells are created by centrifuging at 2,000 rpm for 10 minutes.
- The supernatant is carefully pipetted out and distilled water is added to 9.5-mL mark; the tube is shaken to hemolyze the cells.
- Then 0.5 mL of 17% saline or stock buffer solution is added to restore tonicity. The buffer is prepared by combining NaCl (17.00 g), Na_2HPO_4 (1.13 g), and KH_2PO_4 (0.27 g) and adding distilled water to make 100 mL. This stock solution may be diluted 1 in 20 parts of water and used as a diluent for all regents in subsequent steps.
- Once the standard hemoglobin solution is made, the color standard is prepared from it in tubes with the same diameters that will be used for the test itself. The color standards are made as given in Table 9.1.
- The tubes are then shaken and labeled according to the degree of hemolysis, followed by centrifugation at 2,000 rpm for 10 minutes.

Amboceptor Titration

- Titration of amboceptor should be started from 1:2,000 dilution. The various dilutions (1:2,000 to 1:7,000) of the amboceptor are made up to 10 mL each.
- To the above, an equal volume of the 2% cell suspension is added. The sensitized cells must be prepared 15 minutes before the test.
- A 1:100 dilution of complement is prepared from the pooled guinea pig serum.
- The tubes for the titration of amboceptor are arranged as described in Table 9.2, and titration is performed for each dilution of amboceptor. In this preliminary titration of amboceptor, the complement titration is performed at the same time.

TABLE 9.1
Color Standard Preparation

Percent Hemolysis	0	10	20	30	40	50	60	70	80	90	100
Saline (mL)	0.80	0.80	0.80	0.80	0.80	0.80	0.80	0.80	0.80	0.80	0.80
Standard Hemoglobin (mL)	0.00	0.02	0.04	0.06	0.08	0.10	0.12	0.14	0.16	0.18	0.20
2% RBC (mL)	0.20	0.18	0.16	0.14	0.12	0.10	0.08	0.06	0.04	0.02	0.00

TABLE 9.2
Amboceptor Titration

Tube No.	1	2	3	4	5	6	7	8	9	10
Saline (mL)	0.55	0.50	0.45	0.40	0.35	0.30	0.25	0.20	0.15	0.10
Complement (mL)	0.05	0.10	0.15	0.20	0.25	0.30	0.35	0.40	0.45	0.50
Sensitized cells (mL)	0.40	0.40	0.40	0.40	0.40	0.40	0.40	0.40	0.40	0.40

TABLE 9.3
Amboceptor Titration Reading

Tube no.	1	2	3	4	5	6	7	8	9	10
Amboceptor Dilution										
1:7,000	0	0	0	0	5	10	20	25	35	45
1:6,500	0	0	0	8	20	30	40	55	65	68
1:6,000	0	0	3	10	25	35	50	60	65	70
1:5,500	0	0	5	15	35	45	55	70	80	85
1:5,000	0	0	8	30	50	60	80	85	90	95
1:4,500	0	0	20	40	60	70	85	88	95	98
1:4,000	0	0	25	50	75	80	85	90	95	100
1:3,500	0	10	30	60	80	85	95	100	100	100
1:3,000	0	30	55	80	90	95	98	100	100	100
1:2,500	0	35	60	85	93	98	100	100	100	100
1:2,000	2	37	65	90	95	100	100	100	100	100

- The tubes with the mixed reagents are kept at 37°C in a water bath for 30 minutes followed by centrifugation at 2,000 rpm for 10 minutes.
- Each tube is then compared with the color standard tubes and the results (% hemolysis) recorded as shown in Table 9.3.
- The optimal dilution of amboceptor is that beyond which any increment in amboceptor concentration does not produce any marked change in the quantity of complement required for 50% hemolysis. As we can see from the table, in this instance, the optimum dilution is 1:3,000, and this concentration should be used for daily titration of complement and for sensitization of RBC for the test itself.
- By this method, both the complement and amboceptor titration are performed at the same time.

Complement Titration

The complement should be titrated with the optimal dilution of amboceptor before testing patient serum. After the incubation and centrifugation steps, each tube in the test is compared with the 50% hemolysis tube in the color standard.

Illustrative example: If the tube 3 (containing 0.15 mL of 1:100 dilution of complement) matches the color standard of the 50% hemolysis tube, then it is taken as 1 unit of complement. As 4 units of complement are required in each 0.20 mL, the dilution of complement is made as follows:

- 4 x 0.15 = 0.60 mL of 1:100 dilution of complement.
- Therefore, 0.60/100 = 0.20/x; or x = 33.3.

So, 0.20 mL of a 1:33.3 dilution of complement will be required for the test and for the complement.

Test Procedure

Specimen: Serum sample to be tested.

- The serum from the patient is inactivated for 30 minutes at 56°C.
- Two dilutions of the test should be tested: 1:5 and 1:10.
- The reagents are added as per the template in Table 9.4. Optimal dilution of the specific antigen should be used for the test.

TABLE 9.4
Test Proper and Interpretation

Tube No.	1	2	3	4	5	6	7	8	9	10	11
Saline	—	—	0.30	0.25	0.20	0.30	0.25	0.20	0.30	0.25	0.20
Serum 1:5	0.20	—	0.20	0.20	0.20	—	—	—	—	—	—
Serum 1:10	—	0.20	—	—	—	0.20	0.20	0.20	—	—	—
Complement (4 units)	0.20	0.20	0.10	0.15	0.20	0.10	0.15	0.20	0.10	0.15	0.20
Antigen	0.20	0.20	—	—	—	—	—	—	0.20	0.20	0.20

- After adding the reagents and mixing, the tubes are kept overnight (16–18 hours) at 4°C.
- The next day, 0.40 mL of freshly sensitized sheep RBC is added to each tube, incubated in a water bath at 37°C for 30 minutes, and then read.

Result and interpretation: Interpretation of the test needs careful reading of all the tubes, and the mere presence of hemolysis should not be considered as a negative test result. Unlike the conventional CFT, the test format presented above considers the deterioration of the complement in the presence of the antigen and serum, which takes place during overnight incubation. This deteriorating factor can also be termed *binding power* of the serum or antigen.

As shown in Table 9.4:

- Tubes 3, 4, and 5 contain 2, 3, and 4 units of complement (50%), respectively, to determine the binding power of the 1:5 serum dilution.
- Tubes 6, 7, and 8 contain 2, 3, and 4 units of complement (50%), respectively, to determine the binding power of the 1:10 serum dilution.
- Tubes 9, 10, and 11 show the quantity of complement bound by the antigen.
- The tubes are compared with the color standards containing 30%, 50%, 70%, and 100% hemolyzed cells. Any tube in the test that matches the tube containing 30% hemolyzed cells or higher is considered positive, and those showing less than 30% fixation are taken as negative.
- Positivity in tubes 1 and 2 at 1:5 and 1:10 or only at 1:5 serum dilution with no or negligible binding or anti-complementary effects seen in other tubes is considered a positive test.

Quality assurance: The CFT is a complex test and needs stringent quality control to yield accurate results.

The following controls should be included in each test run:

- Antigen control with no serum (antigen+complement+buffer)
- Complement control with no antigen (complement+ sensitized rbc+buffer)
- Sensitized RBC control (sensitized RBC+ buffer)
- Hemolysis standard control to compare the results (0, 10, 20, ... 100% hemolysis).

Control sera: Both known positive and negative sera.

A test is considered valid if it meets the following criteria:

- Antigen control: 100% hemolysis
- Complement control: 100% hemolysis
- Sensitized RBCs control: 0% hemolysis
- Negative serum control: 100% hemolysis
- The positive control serum is visualized at the expected titer ± one dilution.

Newer Formats of the CFT

Newer formats of the CFT include the following:

Microtiter plate: Instead of using large volumes of the reagents and performing the tests in test tubes, microtiter plates can be used with much smaller volumes of the reagents, making the test more economical. Use of micropipettes considerably eases the procedure. Disposable microplates (96-well round U-bottomed) with lid or cover are recommended.

Use of spectrophotometer instead of color tubes: The use of spectrophotometer to record the optical density obviates the need for preparation of color tubes and prevents the subjective errors of the observers. Otherwise, a plate-reading mirror can be used.

Use of buffer: The veronal buffer is used instead of phosphate-buffered saline.

Commercial Availability of CFT Kits

Since CFT has fallen into disuse in recent times, no ready-made kit is available for any parasitic disease. However, some of the key requirements are available commercially, which can help in devising experimental tests without excessive effort and time being spent in preparation of these biologicals. These include:

Ready-made hemolysin, complement: These are readily available from commercial suppliers (Virion Serion Diagnostica, Germany; Creative Diagnostics, USA; Rockland Immunochemicals, USA; etc.).

Sheep red cells: These are available from various commercial sources (Fisher Scientifics, Merck/Sigma-Aldrich, etc.)

Purified parasite antigen: Purified parasite antigens are available for *Toxoplasma gondii*, various *Plasmodium* species, *Leishmania*, *Acanthamoeba*, *Trypanosoma cruzi*, *Trichomonas vaginalis*, *Cryptosporidium parvum*, *Echinococcus granulosus*, *Schistosoma mansoni*, *Wuchereria bancrofti*, and others (from such suppliers as Virion Serion Diagnostica, Germany, and Creative Diagnostics, USA.)

CONCLUSION

In the past, CFT has served as a reference test for several microbial infections. However, the introduction of easier-to-use test formats has resulted in a steady decline in CFT over the years. It is still available in a few select reference laboratories for select pathogens, but its use has been largely abandoned for routine testing. It remains a highly sensitive test, and the ready availability of sheep RBC, amboceptors, complement, and purified antigens of several parasites from various commercial suppliers can be of help for any investigator to design the test for the diagnosis of a particular parasite.

FURTHER READINGS

Bozicevich J, Hoyem HM, Walston VM. A method for conducting the 50 percent haemolysis end point complement-fixation test for parasitic diseases. *Public Health Rep.* 1946;61:529–534.

Brucellosis Complement Fixation Test (EU Reference Labcold technique) Standard Operating Procedure. Available at: https://sitesv2.anses.fr/en/system/files/private/Brucellosis%20EU-RL%20CFT%20SOP%202010%20Rev5_DRAFT_0.pdf. Accessed October 3, 2024.

Khan LB, Tsai J-YC. A simple, cost-effective undergraduate workshop based on simulated complement fixation test to teach the concept of complement system. *Eur J Health Biology Edu.* 2019;8(1):19–30.

Public Health England. Complement fixation tests. *UK Standards MicrobiolInvest.* 2013b;18(3.2). Available at: http://www.hpa.org.uk/SMI/pdf. Accessed October 21, 2024.

10 Enzyme-Based Tests

INTRODUCTION

The test formats discussed in the previous chapters have relied on the visual interpretation of macroscopic changes occurring during antigen–antibody interaction. The search for more sensitive and specific assays led to the discovery of labeling antibodies to radioisotopes and the development of radioimmunoassay. Health and environmental hazards posed by radioisotopes were a growing concern, however, which paved the way for the use of enzymes as the labeling agent. The initial skepticism that enzymes were too large to be used as labels because of steric interference was soon disproved by the work of Swedish scientists Peter Perlmann and Eva Engvall in 1971. This was followed by the successful development of the first enzyme immunoassay by Organon Research Laboratories in the Netherlands, led by Anton Schuurs and Bauke van Weemen. This technique could detect the analyte at as low as 10^{-6} to 10^{-9} molar concentration.

PRINCIPLE OF ENZYME-BASED TESTS

The enzyme-linked immunosorbent assay (ELISA) can be considered the prototype of all enzyme immunoassays. Such assays can detect the analytes at concentrations less than 1 pg/mL.

There are various formats of ELISA, of which the indirect ELISA for detection of antibody is the one most in use. In it, a solid-phase support is used to first immobilize the antigen against which the antibody in a patient's sample is to be detected. A blocking agent to decrease the probability of nonspecific bonding is commonly applied following antigen immobilization. Next, the sample containing the antibody to be detected is added, and if similar, antigen–antibody bonding will occur. Anti-human antibody labeled with an antigen through covalent bonding is then added to the system, which will bind to the antibody in the antigen–antibody complex already formed. The substrate is then added to system, which will be acted on by the enzyme to produce a colored product. The intensity of the color is then measured in a spectrophotometer (commonly called an ELISA reader), which indicates a positive test reaction. Washing with a mild detergent is important after each step to remove nonspecific or excess unbound proteins without affecting the bound products.

IMPORTANT CONSIDERATIONS IN DESIGNING AN IN-HOUSE ELISA SYSTEM

For developing an in-house ELISA, several factors are important, since they may affect the actual working of the test system.

Test Design Components

a. *The solid-phase support*: Various formats of solid-phase support are available (beads, tubes, or wells). Of these, the 96-well microtiter plates are the most popular, since they allow many samples to be tested simultaneously. These are commonly made of polystyrene, polymethylmethacrylate, or polycarbonate and may have different antigen-binding capacities. It is advisable to use plates with high binding capacity for the proteins that are limited in quantity or expensive. For other purposes, low-capacity binding plates are acceptable, although they may give low absorbance readings. The binding capacity of the microtiter plates can be obtained from manufacturers. Round-bottom wells are preferred in contrast to V- or flat-bottom wells if manual washing is done. If using an automated microwell washer, any configuration is acceptable.

DOI: 10.1201/9781003644224-13

b. *Antigen–antibody coating*: This is the most crucial step in designing an ELISA, since the efficiency of the system depends on it. The common technique is passive adsorption of the protein (antigen or antibody) mediated primarily by hydrophobic interaction, but electrostatic forces also play a role. The protein is adsorbed with the help of an alkaline buffer. The high pH helps dissolve many proteins, and these molecules are protonated with an overall negative charge, which helps when binding to a positively charged plate. Two buffers commonly used include 0.2M carbonate/bicarbonate buffer (pH 9.4) and sodium borate (pH 9.0). The wet plate with the antigen solution can be stored at 4°C up to 4 weeks.

   For longer shelf life, the following procedure may be adopted:

   - Any unbound material is gently washed with the washing solution.
   - A blocking agent is applied to coat the unbound sites on the plate. Sodium casein (1%) or skimmed milk is allowed to act for 30–60 minutes at 37°C followed by another wash. Even though bovine serum albumin (BSA) is commonly used nowadays, sodium casein is considered more effective.
   - 1% glucose or sucrose solution is added and kept for 30 minutes at room temperature.
   - Rapid drying of the plate is done in vacuum or a passing stream of nitrogen, and the plate is kept at 4°C.

c. *Washing solution*: The washing solution for ELISA consists of a mild detergent in a buffered solution. Its purpose is to wash off unbound molecules and to prevent nonspecific bindings. Commonly, 0.05% Tween 20 in phosphate-buffered saline is used as the washing solution. The same solution can also be used to dilute the specimen and antibodies.

d. *Enzyme*: The popular enzymes used as antibody conjugate are horseradish peroxidase (HRP), alkaline phosphatase (AP), and β-galactosidase. HRP is the favored enzyme because, unlike AP, it does not suffer from steric hindrances or high-dose hook phenomenon. However, HRP is incompatible with sodium azide used as a preservative in many biological buffers, because sodium azide can interfere with HRP activity. Enzyme-linked reagents are commercially available for purchase and are recommended for use. However, if the laboratory is using specific monoclonal antibodies, the conjugation step must be performed in-house.

   The two most common methods for coupling are the two-step glutaraldehyde method and the periodate method. For details of these techniques, readers can refer to the monograph by Crowther.

e. *Substrate and chromogen*: Hydrogen peroxide (0.003%–0.015%) is used as the substrate for HRP enzyme while p-nitrophenyl phosphate (pNPP) is the corresponding substrate for AP.

   Different chromogens that can be used for hydrogen peroxide are:

   - Tetramethylbenzidine (TMB) 0.1 mg/mL in 0.05–0.1M phosphate-citrate buffer (pH 5.0).
   - Azino-di-(3-ethylbenthiazoline-6-suiphonic acid) (ABTS) 0.4 mg/mL in 0.05 M phosphate-citrate buffer (pH 4.0).
   - o-Phenylenediamine-HCl (OPD) 0.5 mg/mL in 0.05–0.1 M phosphate-citrate buffer (pH 5.0).

   For AP, the chromogen used is pNPP (1 mg/mL) in 10mM diethanolamine and 0.5mM $MgCl_2$ (pH 9.5). β-Galactosidase requires 70 mg o-nitrophenyl-β-d-galactopyranoside (ONPG) per 100 mL of 0.1 M potassium phosphate buffer (pH 7.0) containing 1 mM magnesium chloride and 0.01 M 2-mercaptoethanol. These chromogens are available in the tablet or powder form from various commercial suppliers such as Sigma-Aldrich and ThermoFisher Scientifics, among others.

   The chromogens are light sensitive, hence the reaction mixture during the ELISA procedure should be incubated in a dark environment. Since ABTS and OPD are possible mutagens, the use of TMB is preferable. Moreover, TMB has the maximum sensitivity to limited amounts of HRP.

f. *Stopping solution*: H_2SO_4 (1M) is used to stop the reaction after the designated time when using HRP. NaOH (1M) or EDTA (0.1M) is used with AP while 2 M Na_2Co_3 is used for β–galactosidase enzyme-based tests.

Types of Antigens and Antibodies

The antigen used for coating the plates can be crude antigen of the parasite or the excretory-secretory product (ESP) of the helminths. One advantage of using crude antigen is that antibodies can be detected against a number of protein fractions of the parasite. Recombinant antigens are being explored to improve the specificity, since the crude antigen results in cross-reactivity with related parasites.

Similarly, antibodies used in ELISA can be monoclonal, polyclonal, or a combination of both types.

a. *Monoclonal antibodies*: These antibodies have high specificity and can be used for all antibody-related steps in any type of ELISA format. In combination with polyclonal antibodies, they can improve the sensitivity and capture a wider array of antigens.
b. *Polyclonal antibodies*: These antibodies are a complex mixture of antibodies against several antigenic epitopes. They have the strength to detect the whole antigen with a higher signal level. Cross-reactivity is again a problem, which can be bypassed using affinity-purified or cross-absorbed polyclonal antibodies. *Matched pairs* are frequently used in sandwich ELISA and consist of two monoclonals or two polyclonals, or a combination of them. It helps capture different epitopes on the same antigenic molecule.
c. *Secondary antibodies*: These antibodies are used with the label (enzyme) and are derived from animal species, hence they are sometimes called anti-human antibodies/globulin. These are raised in various animals such as mouse, rabbit, and goat, among others.

Controls and Cutoff Value

Negative and positive controls are essential prerequisites for performing an ELISA. Samples that contain the analyte of interest are called positive controls. Negative controls are samples that do not contain the analyte of interest and are used to check nonspecific binding. In-house or endogenous positive controls are sometimes used when recombinant antigen is used in the system to validate the functioning of the antigen.

Standard controls are used in the quantitative ELISA for determining the concentration of IgG antibodies in the patient's serum. These are positive controls with predetermined analyte concentrations. A calibration curve of optical densities (OD) is generated using these controls, and analysis of the sample OD in the curve will give the IgG value.

While designing a qualitative ELISA where results are expressed as positive or negative only, calculation of the cutoff value (COV) is essential. The COV is the absorbance value (optical density) above which a sample can be considered positive. For calculations, it is common to use the average absorbance of three known negative samples plus a constant factor. The constant factor is 3–4 standard deviations of the mean absorbance value of more than 100 known negative samples. Another method is to test many known negative samples and to determine the highest absorbance in the distribution curve.

ELISA FORMATS AND THEIR USES

There are four fundamental protocols of ELISA: direct, indirect, sandwich, and competitive (Figure 10.1).

Types of ELISA

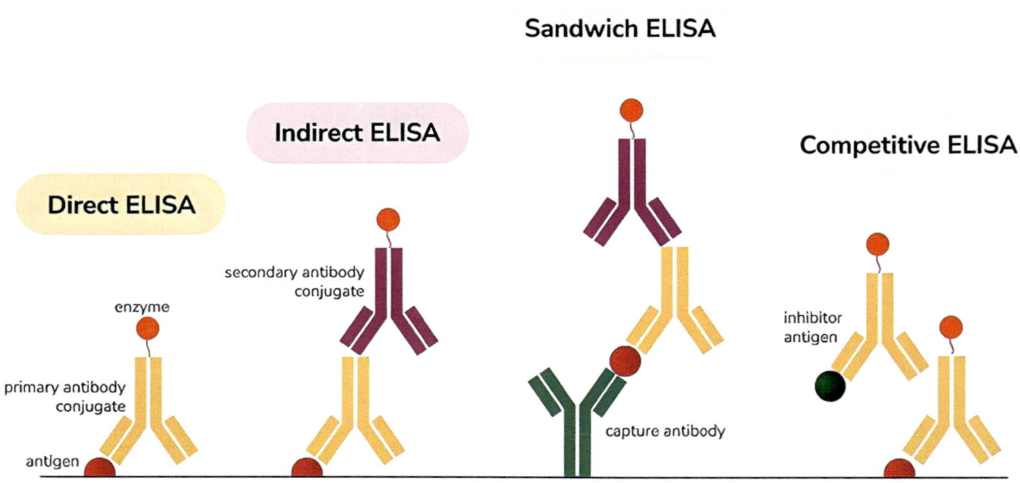

FIGURE 10.1 The various formats of enzyme-linked immunosorbent assay.

(Photo credit: Surmodics IVD, USA.)

Direct ELISA

In this method, the antigen is immobilized on a plate and excess antigen is washed away. Specific antibodies tagged with an enzyme are added, incubated, and washed, after which the substrate-chromogen is added and color development is noted. It is faster than other ELISA methods and requires fewer reagents, resulting in a lower chance of error. However, it can cause higher background noise because all proteins, including the target antigen in the sample, can bind to the plate. It is also less sensitive, since there is no signal amplification. Direct detection, while not widely used in ELISA, is quite common for immunohistochemical staining of tissues and cells.

Indirect ELISA

It is the most commonly used format in which a secondary labeled antibody is used. This format is ideal for determining total antibody concentration in patient samples. The antigen is immobilized on the plate followed by addition of the patient's sample containing antibodies. If a corresponding antibody is present, it will bind to the antigen. The antigen–antibody complex is then detected by using an anti-human secondary labeled antibody.

The anti-human antibody used may be a goat or mouse antibody. The enzyme-conjugated secondary antibody in turn will bind to the primary antibody. The addition of chromogen substrate followed by color development indicates a positive test. Apart from the disadvantage of a longer test duration due to additional incubation steps, this format is superior to direct ELISA because it is more sensitive, more economical (no need for multiple labeled antibodies), and offers greater flexibility.

Sandwich ELISA

This format is suitable for detection of antigens in the patient's sample. The name comes from the fact that the antigen is sandwiched between two antibodies.

In this format, the capture antibody corresponding to the antigen to be tested is attached to the solid support. The sample containing the antigen is then added, followed by a detection antibody. In direct sandwich ELISA, the detection antibody is directly conjugated to an enzyme.If the detection antibody used is unlabeled, a secondary enzyme-conjugated detection antibody is added in a procedure called an indirect sandwich ELISA. This is followed by the usual step of addition of substrate-chromogen system and observing for color development. The sandwich ELISA is two to five times more sensitive than direct or indirect types as well as highly specific, since two antibodies are used to capture the antigen.

While designing an in-house sandwich ELISA, antibody optimization is required to minimize cross-reactivity between the capture and detection antibodies.

The IgM capture ELISA can be considered a type of sandwich ELISA used to detect IgM fraction in the patient's sample. In this format, the microplates are coated with anti-human IgM to bind the corresponding IgM antibodies of the sample. The sample is added followed by a mixture of antigen and enzyme-labeled antibody. The mixture binds to the captured specific IgM antibodies, which is then detected by adding the substrate-chromogen.

Competitive ELISA

Competitive or inhibition ELISA, also known as a blocking ELISA, is the most complex of all the ELISA techniques. It can be adapted for detection of both antigen and antibody. The name comes from the fact that the sample antigen or antibody competes with a reference system for binding to a limited amount of labeled antibody or antigen, respectively. The higher the sample analyte concentration, the weaker the output signals, indicating that the signal output is inversely proportional to the amount of antigen or antibody in the sample.

For antibody detection, the antigen is bound to the plate to which the sample is added. After this, the pretreated dilution of labeled antibody (specific for the solid-phase antigen) is added. This leads to a competition between the labeled and unlabeled antibodies for the binding sites on the antigen. The main advantages include no need for sample processing, greater robustness and consistency, and high flexibility, as it can be based on any of the other three formats (direct, indirect, or sandwich). It is commonly used when only one antibody is available for the antigen of interest. It is also useful for detection of minute quantities of antigens that cannot be bound by two different antibodies like in the sandwich ELISA.

This section presents four in-house design and optimization methods for various formats of ELISA in diagnosis of different parasites.

Designing and Optimizing ELISA for the Serum Antigen Detection in Strongyloidiasis

To date, there are no commercial kits available for detection of *Strongyloides* antigen in stool or serum samples, although there are quite a few reports of in-house ELISA designs. The earlier studies used polyclonal antibodies raised in rabbits against antigens from different species such as *Strongyloides stercoralis*, *Strongyloides ratti*, and *Strongyloides venezulensis* for the detection of coproantigens. Subsequently, monoclonal antibodies produced by the hybridoma technique were used. Recent technological developments have helped in production of recombinant monoclonal antibodies and single-chain fragment variable (scFV) antibodies. These antibodies have been used for detection of some parasitic diseases such as neurocysticercosis and schistosomiasis. Here, we describe an ELISA for detecting *S. stercoralis* antigen in serum samples by Balachandra et al. where scFV recombinant antibody has been used.

Reagents and Equipment Required

Reagents: bovine serum albumin; skim milk; PBS; Tween-20; carbonate-bicarbonate buffer; azino-bis-ethylbenzothiazoline-sulfonic acid (ABTS); HRP.

Equipment and labware: surface plasmon resonance (SPR) Biacore™ x100 instrument; microtiter plate; automated micropipettes; ELISA reader.

Others: monoclonal antibody; T7 Express *Escherichia coli*; mouse anti-human IgG Fc conjugated to HRP; goat anti-human IgG F(ab')2-HRP.

Design and Optimization of the Test

Four rMAb proteins representing unique antibody gene family (rMAb5:IgHV3-LV1; rMAb6: IgHV3-LV6;rMAb14:IgHV5-KV3; rMAb23:IgHV3-LV3) were expressed in SHuffle® T7 Express *Escherichia coli* cells.

- Surface plasmon resonance analysis is done to select the best scFv recombinant monoclonal antibody for ELISA.
- Two types of antibody conjugate (mouse anti-human IgG Fc conjugated to HRP and goat anti-human IgG F(ab')2-HRP), four concentrations of antibody coated on the wells (5, 10, 20, and 40 µg/mL), two blocking reagents (3% bovine serum albumin and 2% skim milk), two serum dilutions (1:100 and 1:200), and two conjugate dilutions (1:1,000 and 1:2,000) are evaluated for optimum results.
- rMAb23 showed the best response with regards to high association rate constant, low dissociation rate constant, and low equilibrium dissociation constant.
- The following parameters produced optimum results: 10 µg/mL rMAb23, 2% skim milk diluted in 0.05% Tween-20 in phosphate-buffered saline (PBS-T), 1:100 serum dilution, and 1:1,000 goat anti-human IgG F(ab')2-HRP.
- The ROC curve showed a cutoff value (COV) of 0.5002 at a reading taken at 405nm wavelength.

Test Procedure

- ELISA plate wells are coated with scFv rMAb diluted in carbonate-bicarbonate coating buffer (pH 9.6) and incubated overnight at 4°C.
- The wells are washed three times for 3 minutes each with PBS-T at 37°C and then centrifuged at 800 rpm.
- Blocking is done with 2% skim milk, after which the wells are incubated for 1 hour at 37°C, centrifuged at 800 rpm, and then washed.
- The serum samples are diluted (1:100) in PBS and added to the wells and incubated for 2 hours at 37°C, centrifuged at 300 rpm, and then washed.
- The 1:1,000 dilution of goat anti-human IgG F(ab')2-HRPdiluted in 2% skim milk (100 µL/well) is added and incubated for 30 minutes at 37°C, centrifuged at 300 rpm, and washed.
- 1 mg/mL ABTS substrate is added, and the absorbance is recorded at 405 nm.

Results and interpretation: In a positive test, a blue-green color develops (Figure 10.2), while a negative test produces no color. A cutoff optical density value of 0.5002 was used for interpretation of results. Any test well showing OD above this value is considered positive.

Quality assurance: Three pooled serum samples from known *Strongyloides*-positive patients were used as positive controls. It included patients with IgG and IgG4 positivity with positive stool microscopy or stool PCR.

Three pooled serum samples from *Stongyloides*-negative cases were used as negative controls. These included healthy donors as well as patients diagnosed with other parasitic infections including amoebiasis, toxoplasmosis, schistosomiasis, hookworm, and toxocariasis.

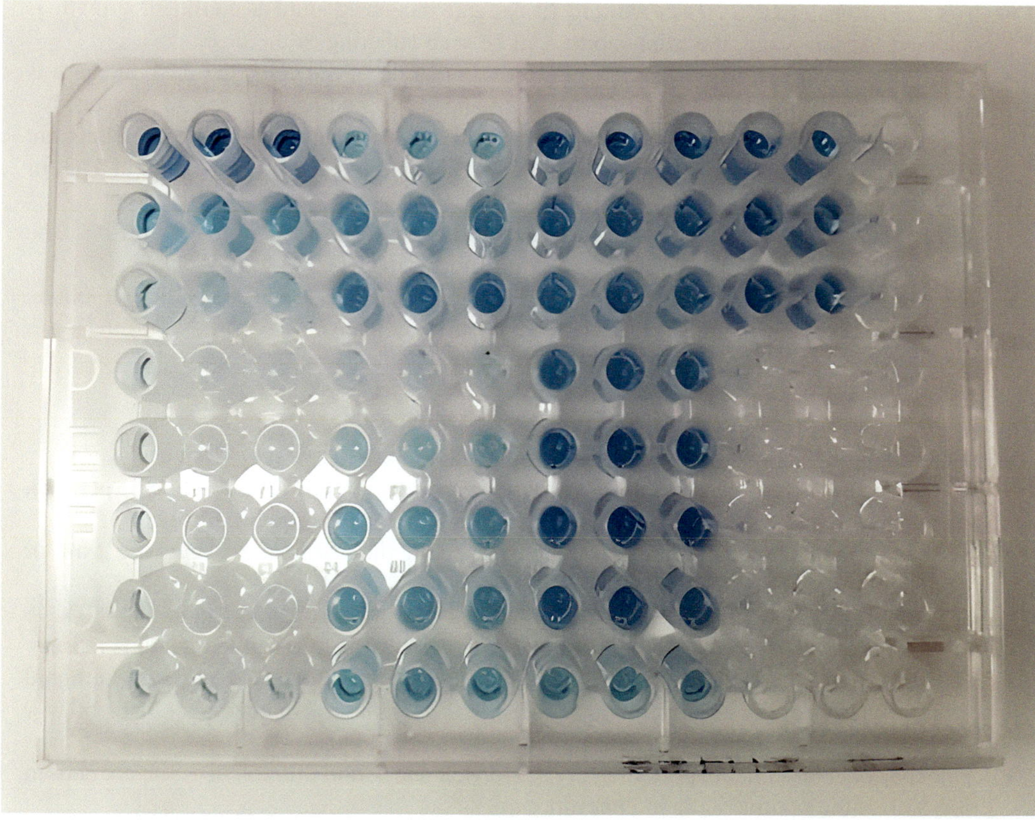

FIGURE 10.2 An ELISA microtiterplate with horseradish peroxidase–linked secondary antibody. Blue color development signifies a positive test.

(By Ajpolino – Own work, CC0, https://commons.wikimedia.org/w/index.php?curid=88819005.)

Designing and Optimizing ELISA for the Serum Antibody Detection in *Acanthamoeba*

Antibody detection is useful for diagnosis of *Acanthamoeba* encephalitis, pulmonary infections, as well as for seroprevalence studies in the population. The given protocol has been adapted from the study by Chappell et al.

Reagents and Equipment Required

Reagents: peptone-yeast extract glucose medium; fetal bovine serum; Eagle minimum essential medium; PBS (pH 7.2; 7.6); formalin solution; sodium carbonate buffer (pH 9.6); skim milk; 2,2′-azino-di-(3-ethylbenzthiazolinesulfonate); HRP; Tween-20; liquid nitrogen.

Equipment and labware: tissue culture flasks; ice bath; centrifuge; hemocytometer; tissue homogenizer; ultrasonicator; microtiter plate; spectrophotometer/ELISA reader.

Others: *Acanthamoeba polyphaga* trophozoites; New Zealand White rabbits; HRP-conjugated goat anti-human IgG.

Design and Optimization of the Test

Antigen Preparation

- *Acanthamoeba polyphaga* is grown in tubes containing a peptone-yeast extract glucose medium supplemented with 5% heat-inactivated fetal bovine serum and a minimal essential medium vitamin mixture at room temperature for 7–10 days.

- It is then transferred to tissue culture flasks and incubated until a confluent growth is obtained. At this stage, the trophozoites can be seen adhering to the flask.
- To dislodge the trophozoites and to induce the rounding of the cells, the flask medium is replaced with 50 mL of 0.15 M PBS (pH 7.6) and kept in an ice bath for 30 minutes.
- The flasks are then centrifuged at 4,500 rpm for 10 minutes, and the cell pellets are resuspended in 10 mL of PBS with 1% formalin to fix the cells. It can be kept at 4°C for 1 week.
- Before use, the cells are washed three times in PBS and resuspended in 10 mL PBS. The number of amoebic cells is counted using a hemocytometer.
- The necessary dilution is then done in 0.05 M sodium carbonate buffer (pH 9.6) to obtain the required number of cells/mL for antigen coating on plates.

Production of Hyperimmune Serum and Assay Optimization

- Hyperimmune serum is raised in New Zealand White rabbits (weight, 2 kg each).
- Antigen is prepared by using 2 x 10^5 *A. polyphaga* cells per mL in 0.15 M PBS (pH 7.2), after which they are subjected to four freeze–thaw cycles using liquid nitrogen.
- Weekly immunization for 4 weeks is done by injecting the marginal ear vein with 2 mL of the antigen preparation.
- At the end of fifth week, the animals are bled and serum collected and stored at –80°C.
- Negative control serum is obtained from the animals bled before antigen administration.
- To ascertain the optimum antigen concentration to be used on the plate, checkerboard dilution with twofold serial dilutions of fixed *A. polyphaga* trophozoites (24 to 50,000 trophozoites) and 1:100 to 1:12,800 rabbit antiserum is done.
- Amoeba concentration of approximately 3,000/well and serum dilution of 1:400 are chosen to establish the optimal antigen concentration.
- Pre-immune rabbit serum is tested in the same fashion on a separate plate.
- In a further step to find out whether the intact organism is superior as an antigen compared with disrupted cells, the same number of organisms (3,000 trophozoites) are homogenized and sonicated.
- The cytosolic antigens show higher absorbance and hence may contribute to nonspecific reactivity and a high estimation of seroprevalence.
- Thus, the intact trophozoites are selected for the assay proper.
- The COV is calculated by determining the mean absorbance of nonreactive samples added to 3 standard deviations (mean+3SD).

Test Procedure

- A known number of fixed amoebae are placed in microtiter wells and incubated at 37°C for 1 hour, followed by overnight incubation at 4°C.
- Wells are blocked for 90 minutes at 37°C with 200 mL of 5% dry milk-PBS.
- Human sera (1:10 dilution, 100μL) are added to wells, incubated at 37°C for 1 hour, and then washed.
- Commercially obtained HRP-conjugated goat anti-human IgG (1:1,000) is added.
- This is followed by addition of peroxidase-activated (final concentration, 0.03%) 2,2`-azino-di-[3-ethylbenzthiazolinesulfonate].
- The absorbance is read spectrophotometrically at 414 nm after 3, 5, 10, 15, and 30 minutes.
- 0.15 M PBS (pH 7.2), containing 0.1% Tween-20, is used as the washing solution at each step.

Results and interpretation: Sera yielding optical densities of >0.25 are used to define the cutoff for positive results.

Quality assurance: Known positive and negative controls should be run with each test run.

Design and Optimization of Sandwich ELISA for the Detection of Hydatid Antigen in Urine for Diagnosis of Cystic Echinococcosis

The diagnosis of CE primarily relies on imaging methods and serological tests. Imaging methods are costly or not available in most of the endemic areas. Thus, serodiagnosis plays an important role, as it helps in the evaluation of radiologically unclear cases and the identification of the patient's immune status. Serodiagnosis by demonstration of serum antibodies does not confirm clinical diagnosis because of nonspecific reactions and persistence of anti-hydatid antibodies for several years, even after surgical removal of a hydatid cyst. Hence, diagnostic methods based on the detection of parasite antigens are more useful, as they help measure parasite burden more directly, giving an indication of the activity and intensity of infection as well as assessing medical and chemotherapeutic treatment of CE.

The collection of blood for demonstration of antigen is an invasive procedure requiring technical expertise and disposable syringes to prevent the risk of transmission of serious infectious agents such as HIV and hepatitis B virus. Urine collection, by contrast, is a noninvasive procedure and considered user-friendly in obtaining a diagnosis. Hydatid antigen detection for diagnosis of CE was first reported by Parija et al. in 1997 by CIEP using hyperimmune antisera raised against crude human hydatid cyst fluid. Subsequently, Co-A was also used to detect the urinary hydatid antigen by Parija and associates.

The protocol for the development of in-house sandwich ELISA presented here, optimized for hydatid antigen detection in urine, has been adapted from the study by Chaya and Parija.

In this format, a capture antibody (affinity purified polyclonal hydatid antibodies) is coated onto a microtiter plate to which the sample containing the target antigen (hydatid antigen) is added. Subsequently, a detector antibody (rabbit anti-hydatid IgG conjugated to HRP) is added, followed by the substrate (OPD), resulting in color development.

Reagents and Equipment Required

Reagents: phosphate-buffered saline (PBS) (pH 7.2); washing buffer (sterile PBS [pH 7.2] containing 0.1% Tween-20 [PBS-T]); PBS (pH 7.2) containing 2% bovine serum albumin (BSA); ortho-phenylenediamine (OPD); stop solution (50 μL of 2 M H_2SO_4 [pH 9.6]); sodium metaperiodate; phenylmethylsulphonyl fluoride.

Equipment and labware: ELISA microplate washer and reader; micropipettes (range: 200–1,000 μL; 20–200 μL, 1–20 μL); disposable microtips (sizes: 200–1,000 μL; 20–200 μL, 1–20 μL); incubator; plate sealer; 96-well flat-bottomed microtiter ELISA plates; dialyzer.

Others: affinity purified polyclonal hydatid antibodies; rabbit anti–hydatid IgG conjugated to HRP; urine sample from surgically confirmed and ultrasound proven CE; disease controls (amoebiasis, filariasis, neurocysticercosis and toxoplasmosis cases); healthy controls.

Key Considerations

Antigen Preparation

i *Antigen source*:

- Surgically removed human hydatid cyst are collected in a sterile container.
- The cysts are washed in several changes with sterile PBS (pH 7.2).
- A part of the cyst content is preserved in a 10% solution of formalin for histological section to observe cyst wall morphology. The presence of a laminated layer, a unique feature of hydatid cysts; along with the germinal layer and hooklets, demonstrated by hematoxylin and eosin (H&E) staining, helps confirm the hydatid ethology of the cyst.
- The hydatid fluid is aspirated from the cyst using a sterile syringe and collected in sterile centrifuge tube.
- The fluid is centrifuged at 2,000 rpm for 30 minutes at 4°C.
- The clear supernatant fluid is collected as hydatid fluid.
- The fluid is checked for sterility by inoculating it on blood agar and Mac Conkey agar and incubated at 37°C overnight.
- The sediment containing the protoscoleces is also collected.

Checking the Fertility of Hydatid Cyst

- The fertility of the cyst is detected by viability tests such as flame cell activity and vital staining.
- Flame cell activity is demonstrated by making a wet mount preparation of the cystic fluid on a slide and examining under 10x and 40x magnification.
- The viability of the protoscoleces is evidenced by faint flickering movements.
- Vital staining is carried out by adding a few drops of 0.1% aqueous eosin to a drop of hydatid fluid on a slide.
- Viable protoscoleces do not take up the stain, whereas the dead protoscoleces are readily stained with the eosin.

Preparation of Somatic Hydatid Antigens

Four different somatic antigens are obtained from the same cysts for immunodiagnosis of CE and the preparation as follows:

i *Hydatid cyst fluid (HCF) antigen*
- The fluid is aspirated aseptically from the fertile hydatid cyst and is collected in a sterile centrifuge tube.
- The fluid is centrifuged at 2,000 rpm for 30 minutes at 4°C and dialyzed extensively against distilled water.
- To this, 0.5M of phenylmethylsulphonyl fluoride (PMSF) is added at a proportion of 1/100 and stored at −20°C.

ii *Hydatid protoscolex (HPR) antigen*
- The sediment containing the protoscoleces obtained after centrifugation of the cyst fluid at 2,000 rpm for 30 minutes are collected in a sterile tube.
- The protoscoleces are washed three times with PBS (pH 7.2).
- The washed protoscoleces are then subjected to three cycles of freezing and thawing and resuspended in 10 times the volume of PBS (pH 7.2) containing 0.5 M PMSF (at a ratio of 1 in 100).
- The protoscolex suspension is then sonicated on ice in a 150-W ultrasonic disintegrator, in 10-second bursts with 5-second intervals, until no intact protoscoleces are visible microscopically.
- The sonicate is left at 4°C overnight and then centrifuged at 10,000 rpm for 30 minutes.
- The supernatant is collected, dispensed in small aliquots, and stored as protoscolex antigen at −20°C.

iii *Hydatid cyst wall (HCW) antigen*
- After aspiration of cyst fluid, the collapsed cyst membranes are washed three times in PBS (pH 7.2) and homogenized using mortar and pestle.
- The homogenized tissue is collected in a sterile container and kept at −20°C overnight.
- It is then thawed and sonicated on ice in a 150-W ultrasonic disintegrator, in 60-second bursts with 5-second intervals, until no intact tissue bits are visible to the naked eye.
- 0.5M of PMSF (ratio 1:100) is added to the sonicated suspension, left at 4°C overnight, and then centrifuged at 10,000rpm for 30 minutes.
- The supernatant is collected, dispensed in small aliquots, and stored as cyst wall antigen at −20°C.

iv *Complete homogenate hydatid (CHH) antigen*

The complete homogenate hydatid antigen is prepared by mixing equal volumes of hydatid cyst fluid antigen, hydatid protoscolex antigen, and hydatid cyst wall antigen.

Protein content of each antigen preparation is estimated using the standard Lowry method, readings are taken at 660 nm in a UV-visible spectrophotometer (Cintra 5, Australia), after which the mixture is aliquoted into 1-mL cryoprotected vials and stored at –20°C.

Antibody Preparation

Either polyclonal or monoclonal antibodies or both are used as capture and detector antibodies for the sandwich ELISA for detection of hydatid antigen in the sample for diagnosis of CE.

Preparation of Polyclonal Hydatid Antibodies

- Two adult rabbits (3–5 kg each) are bled 1 day before to start immunization.
- This serum is used as the normal rabbit serum control.
- For immunization, the emulsion is prepared by mixing equal volumes of complete homogenate hydatid (CHH) antigen and Freund complete adjuvant.
- 0.5 mL of this emulsion is injected intramuscularly into all four limbs of each rabbit.
- After 6 weeks, a booster dose of 0.5 mL of the same antigen preparation with Freund incomplete adjuvant is administered.
- Blood for the test is obtained from the ear vein of each rabbit 10 days after the booster. Approximately 0.5 mL of blood is collected from each rabbit.
- The level of antibody raised in rabbit is tested by indirect hemagglutination using the method of Parija et al.
- Once the antibody titer is significant (1:1,024 and above) in test blood, another collection is done to begin purification of antibody.

Preparation of Affinity Purified Polyclonal Hydatid Antibodies

Chromatographic purification of antibody (IgG) from rabbit sera is used in the test.

- Purification of hyperimmune sera is carried out by pooling high-titer immune sera, and the IgG fraction is purified by using commercially available protein A-Sepharose.
- Briefly, the antibody solution (100 µg/500 µL) is passed through 1 mL of protein A-Sepharose column.
- Different fractions are collected at different intervals using separate sterile containers.
- The detection of protein in collected fractions is tested by Coomassie brilliant blue G-250. This is done by using a U-bottomed microtiter plate and adding equal drops (20 µL) of eluted different fractions separately and Coomassie brilliant blue G-250 dye.
- The presence of protein is indicated by the blue color; in the absence of protein, no coloring develops.
- Different fractions are concentrated using the ammonium sulphate precipitation method.
- The precipitate is collected by centrifugation at 3,000 rpm at 4°C in a cooling centrifuge for 30 minutes.
- The precipitate is then dialyzed against PBS (pH 7.2) at 4°C overnight. The clear dialysate is stored at −20°C until use.

Confirming the Immunoreactivity of Affinity Purified IgG Fractions of Rabbit Hyperimmune Sera Raised Against Complete Homogenate Hydatid Antigen

EITB is performed using a crude extract of hydatid cyst wall antigen, with different fractions of purified rabbit anti-hydatid IgG used to observe the immunoreactivity of affinity purified IgG.

Conjugate Preparation

Coupling of rabbit anti-hydatid IgG with HRP is carried as follows:

- Two milliliters of rabbit anti-hydatid IgG is dialyzed overnight against sodium carbonate bicarbonate buffer (pH 9.5).

- A fresh 0.2M solution of sodium metaperiodate ($NaIO_4$) is prepared.
- 0.1 mL of this solution is added to 0.5 mL solution of HRP (2 mg in 0.5 mL double-distilled water). The mixture immediately turns green.
- This mixture is kept in a shaker at room temperature for 20 minutes.
- It is then dialyzed against 1.0mM sodium acetate buffer (pH 4.4) at 4°C for 5 hours with four to five changes.
- The pH is raised to 9.0–9.5 by adding 10 µL of 0.2 M sodium carbonate bicarbonate buffer (pH 9.5), and immediately 2 mL of rabbit anti-hydatid IgG is added.
- This mixture is kept in a shaker for 2 hours at room temperature.
- 100 µL of freshly prepared sodium borohydride solution is added and kept at 4°C. This conjugate is dialyzed overnight against 0.1M borate buffer (pH 7.4) at 4°C with two to three changes.
- Finally, BSA at a final concentration of 5 mg/mL is added, distributed into vials, and stored at −20°C.

Determination of Optimal Dilution

- A flat-bottomed microtiter plate is coated with a known concentration of 5 µg/well (100-µL volume) of cyst wall antigen in PBS (pH 7.2) per well and kept overnight at 4°C.
- The plates are washed with PBS-Tween-20, and nonspecific sites are blocked with a blocking solution (PBS pH 7.2 with 2% BSA) for 3 hours at 37°C.
- The plates are washed three times with PBS-Tween-20.
- A 100 µL of optimally diluted positive and negative serum (1:1,000 in PBS-Tween-20) is added to each well and incubated at 37°C for 90 minutes.
- After washing, 100 µL of doubling dilution (100, 200, 400, 800, 1,600, 2,400) of rabbit anti-hydatid IgG–HRP conjugate diluted in PBS-Tween-20 is added and incubated at 37°C for 1 hour.
- Final washing is followed by addition of 100 µL of OPD substrate and incubated at room temperature in the dark.
- The reaction is stopped by adding 50 µL of 2 M sulfuric acid.
- The OD is read using an ELISA reader at 450 nm.
- The conjugate in the highest dilution that generated the maximum difference between antibodies-coated positive and negative control wells is considered the optimal dilution.

Specimen Processing: Fifty to 100 mL of urine samples are collected from each of the patients, as well as controls, in sterile glass vials using aseptic techniques.

Concentration of Urine

- The entire volume of 50–100 mL of urine sample is mixed with 35–70 g of ammonium sulphate (at a ratio of 70 g ammonium sulphate to 100 mL of urine) to obtain 100% (w/*v*) saturation.
- The mixture of urine and ammonium sulphate is centrifuged at 3,000 rpm at 4°C in a cooling centrifuge for 30 minutes.
- The precipitate is dialyzed against PBS pH 7.2 at 4°C overnight.
- The clear dialysate is stored as concentrated urine specimen at −20°C until use.

Sandwich ELISA Optimization

- Optimum concentration of coating antibody (affinity purified polyclonal antibodies) and sample urine are determined by a checkerboard titration curve.
- The sensitivity of the affinity purified polyclonal antibodies-based sandwich ELISA is determined using serial dilutions (from 10 to 5,000 ng/mL) of artificially added cyst wall antigens in healthy urine.

- The cutoff value for sandwich ELISA is determined by receiver operating characteristic (ROC) curve analysis.
- The ROC curves are generated by varying the cutoff point for a positive test result and plotting the true-positive rate (TPR) (sensitivity) against the false-positive rate (FPR) (specificity) for each cutoff.
- Curves that are closer to the ideal point (TPR = 1, FPR = 0)—that is, nearer the upper-left corner of the ROC plot—are superior to curves positioned lower and toward the right.

Test Procedure

- Capturing antibody coating: Polyvinyl high binding microtiter ELISA plates (NUNC, New Zealand) are coated with 100 µL (5 µg/well in PBS [pH 7.2]) of the affinity purified polyclonal hydatid antibodies (complete homogenate hydatid [CHH]antigen) in PBS (pH 7.2) and then incubated overnight at 4°C undisturbed.
- The un-adsorbed antibodies are removed by washing the plates with a washing buffer (sterile PBS [pH 7.2] containing 0.1% Tween-20 [PBS-T]).
- The uncoated reactive sites in the wells are blocked by PBS (pH 7.2) containing 2% BSA by incubating for 3 hours at 37°C.
- Plates are washed three times with PBS-T as before.
- 100 µL of urine is added and incubated for 1.5 hours at 37°C.
- The plates are washed three times with PBS-T as before to remove unbound materials from the sample.
- Secondary antibody (conjugate) (rabbit anti-hydatid IgG conjugated to HRP [in-house preparation]) is used at a dilution of 1:400 with PBS (pH 7.2) containing Tween-20 (0.05%), and 100 µL is dispensed to all the wells and incubated for 1 hour at 37°C in the dark.
- Plates are washed three times with PBS-T as mentioned before to remove unbound conjugate.
- The substrate solution is prepared by adding 6 mg of OPD in 10 mL of PBS (pH 7.2) containing 0.05% Tween-20 and 10 µL H_2O_2.
- 100 µL of the substrate solution per well is dispensed to all the wells and incubated for 15–20 minutes at 37°C in the dark to develop optimum color.
- The reaction is stopped by adding 50 µL of 2 M H_2SO_4 per well.
- The absorbance is recorded at 450 nm using the ELISA reader.

  Results and interpretation: Test results were evaluated based on the cutoff value determined earlier by the ROC curve. Any OD value above the cutoff is considered a positive test.

  Quality assurance: Urine samples from surgically confirmed and ultrasound proven CE controls is used as positive controls. Negative controls include urine sample from other infections (amoebiasis, filariasis, neurocysticercosis, and toxoplasmosis cases) and healthy individuals.

Design and Optimization of Dot-ELISA for the Detection of Hydatid Antigen in Urine for Diagnosis of Cystic Echinococcosis

The principle of Dot-ELISA is similar to that of sandwich ELISA, except it uses nitrocellulose membrane (NCM) strips instead of microtiter plate. NCM are dotted with affinity purified hydatid antibodies, to which a sample containing hydatid antigen is added. Subsequently, the addition of the detector antibody (rabbit anti-hydatid IgG conjugated to horse radish peroxidase enzyme) followed by substrate 3,3' diaminobenzidine (DAB) leads to a colored spot.

The given protocol for the development of in-house Dot-ELISA, optimized for hydatid antigen detection in urine, has been adapted from the study by Swarna and Parija.

Reagents and Equipment Required

Reagents: nitrocellulose membrane (NCM); PBS (pH 7.2); washing buffer (sterile PBS [pH 7.2] containing 0.1% Tween-20 [PBS-T]); PBS (pH 7.2) containing 2% BSA; 3,3'diaminobenzidine (DAB); stop solution (50µL of 2M H_2SO_4).

Equipment requirements: micropipettes (range: 200–1,000 µL; 20–200 µL, 1–20 µL); plastic strips; plate sealer; disposable microtips (sizes: 200–1000 µL; 20–200 µL, 1–20 µL); rocker shaker; refrigerator.

Others: affinity purified polyclonal hydatid antibodies; rabbit anti-hydatid IgG conjugated to HRP.

Dot-ELISA Optimization

The following Dot-ELISA procedure is optimized for the diagnosis of cystic echinococcosis.

Optimization of Dot-ELISA conditions is performed to know the concentration of coating antibody to be dotted and the dilution of urine used. The optimum concentration of detecting antibody dilution (i.e., purified rabbit anti-hydatid IgG coupled to HRP—an in-house preparation) based Dot-ELISA is determined using serial dilutions (from 10 to 5,000 ng/mL) of artificially added cyst wall antigens in healthy urine.

Development of the Test

- A NCM of 0.5 x 0.5 cm in size is cut and mounted onto a plastic strip.
- A volume of 2 µL (5 µg/strip) of affinity purified polyclonal hydatid antibodies is dotted onto individual strips and air dried for 30 minutes.
- The un-adsorbed antibodies are removed by washing the strips with the washing buffer.
- The uncoated reactive sites in the strips are blocked by PBS (pH 7.2) containing 2% BSA by incubating for 3 hours at 37°C in a rocker shaker.
- Strips are washed three times with PBS-T.

Test Procedure

- One milliliter of concentrated urine diluted at an optimal dilution of 1:5 is prepared in PBS-T, and the strips are incubated for 1.5 hours at 37°C with constant shaking.
- The strips are washed three times with PBS-T as before to remove unbound antibodies in the sample serum.
- Secondary antibody (conjugate) incubation is performed by using rabbit anti-hydatid IgG conjugated to HRP (in-house preparation). It is used at a dilution of 1:400 with PBS (pH 7.2) containing Tween-20 (0.1%), and the strips are incubated for 1 hour at 37°C in the dark with constant shaking.
- Strips are washed three times with PBS-T as before to remove unbound conjugate.
- Substrate solution is prepared by adding 3 mg of DAB (Sigma, USA) in 5 mL of PBS (pH 7.2) containing 0.1% Tween-20 and 5 µL H_2O_2.
- Substrate solution is added to the strips and incubated for 15–20 minutes at 37°C in the dark under constant rocking to develop color.
- The reaction is stopped by washing with double-distilled water. The strips are air dried and stored in the dark.

  Results and interpretation: The positive reaction is indicated by the appearance of a brown spot in 2–5 minutes. The negative reaction is indicated by the absence of any colored spot in 2–5 minutes.

  Quality assurance: A urine sample from surgically confirmed and ultrasound-proven CE controls is used as positive control. Negative controls include a urine sample from other infections (amoebiasis, filariasis, neurocysticercosis, and toxoplasmosis cases) and healthy controls.

Quality Assurance for all ELISA Tests (Commercial and Experimental)

Internal quality control (IQC): It must be in place to ensure that the test results are within the limits of precision and accuracy. All parameters that can influence the test performance need monitoring, including:

- Equipment performance (plates, pipettes, automatic washers, spectrophotometer)
- Stability of reagents and their shelf life
- Sample preparation and quality
- Operator performance
- The use of QC samples should be run at regular intervals. This can control and validate within-assay and between-assay bias, drift, and shift in results.

External quality assurance (EQA): It helps in providing assessment of a test performance and complements the IQA. A third-party organization sends the same sample to several participating laboratories performing the same test at regular intervals. These samples are tested routinely along with other samples without any special consideration, and results are sent back to the organizer to find out the performance accuracy of a particular laboratory.

Operator performance and chart: Individuals performing the test should monitor the performance of the tests. This is important when several operators may be doing the same test (e.g., different shifts). This helps produce a unified approach to control the test performance and results and identify any discrepancies in day-to-day performance. The performance status of the test can be viewed by anyone to evaluate the standard of the laboratory. This can be done effectively by using charts. Either daily detailed data charts to plot the data from individual ELISA plates or a summary data chart to plot data from all plates used on any day can be used. The charts should be reviewed regularly and results discussed with the operators by the laboratory manager or QC staff in case of any deviation, so that appropriate action can be taken.

Table 10.1 shows the common problems encountered in ELISA with their possible causes and remedial measures.

Commercially Available Kits

The ELISA technique has been used for parasitological diagnosis in the form of both antigen and antibody detection. Table 10.2 shows the diagnostic kits available commercially for different classes of parasites. Details of the suppliers can be found in Appendix B. In addition, several in-house tests for antigen or antibody detection of parasitic infections have been described and validated.

TABLE 10.1
Troubleshooting in ELISA Performance

Problem	Possible Causes	Possible Solutions
No signal	Incorrect assay set up or incorrect reagents used	Review protocol, repeat assay using a positive control
	Incorrect secondary antibody used	Repeat assay using the correct secondary antibody
	Not enough antibody used	Increase the concentration of the primary and/or secondary antibody
	Detection reagent too old or contaminated	Use fresh reagents
	Antigen not coated properly	Longer coating times, different coating buffers to be used
	Plate reader has the wrong settings	Check reader for wavelength, filters
	Antibody stored at 4°C for several weeks or undergone repeated freeze–thaw cycles	Use fresh antibody that has been stored at −20°C or below

(Continued)

TABLE 10.1 (Continued)

Problem	Possible Causes	Possible Solutions
Weak signal	Insufficient amount of antigen coated onto microtiter plate	Use more antigen for coating or change coating buffer
	Not enough antibody used	Optimize antibody concentrations for the assay; higher concentration may be needed
	Detection reagent too old, contaminated, or used at the wrong pH	Fresh detection reagents at the correct pH must be used
	Detection reagent too diluted	Higher concentration of detection reagent may be needed
	Plate reader has the wrong settings	Reader should be at correct settings for wavelength and filters
	Incubation temperature too low	Optimize incubation temperature; reagents should be brought to room temperature before use
High background signal (noise)	High concentration of secondary antibody	Reduce concentration by diluting 1:10
	No absorption of antigen/antibody to wells	Do overnight absorption at 4°C
	Plate not clean	Use new clean plate
	Nonspecific antibody binding	Suitable blocking buffer or an affinity purified antibody to be used
	Too few washings cycle	Increase the washing cycles
	Wrong concentration of blocking agent	Use correct concentration
	Incubation with substrate carried out in the light	To be performed in the dark
	Evaporation of solution from wells during incubation	Cover the plates during incubation
	Pipetting errors	Calibrate the pipettes.
Slow color development	Wrong incubation temperature	Assay to be done at room temperature or as per protocol
	Contaminated solutions	Use fresh solution
	Detection reagent too old, contaminated, or used at the wrong pH	Use fresh reagents at appropriate pH

TABLE 10.2
Available Commercial ELISA Kits for Some Important Parasitic Infections

	Antigen Detection	Antibody Detection
Protozoa	*Entamoeba histolytica* (stool) *Giardia intestinalis* (stool) *Cryptosporidium* spp. (stool) *Plasmodium* spp. (blood) *Leishmania* spp. (urine) *Trichomonas vaginalis*	*Entamoeba histolytica* *Toxoplasma gondii* *Trypanosoma cruzi* *Trichomonas vaginalis* *Leishmania* spp. *Cryptosporidium* spp. *Giardia intestinalis*
Nematode	*Wuchereria bancrofti* (blood)	*Strongyloides stercoralis* *Trichinella* spp. *Toxocara canis* *Ascaris* spp. *Anisakidae*
Cestode	Cysticercosis (*Taenia solium*) (blood)	*Taenia solium* *Echinococcus granulosus* *Echinococcus multilocularis*
Trematode		*Paragonimus westermanii* *Schistosoma* spp. *Fasciola hepatica*

Note: For the full list and suppliers of these kits, please refer to Appendix B.

CONCLUSION

Designing an in-house ELISA is a technically demanding job. Careful consideration of multiple parameters is required, especially the optimal antigen and antibody concentrations, which are established using checkerboard titrations. The quality of reagents is equally important, and it is advisable to obtain them from reliable suppliers. Labeled secondary antibodies and coated plates can be obtained commercially. The test results must be validated, with large numbers of positive and negative samples and assays must deliver consistent and reproducible performance. Even though commercially available assay kits are always preferable, they are currently available for only a limited number of parasites. Reports are available for many in-house ELISA tests both for antigen and antibody detection. The commercial availability of these assays, through collaboration between research institutes and kit manufacturers or via technology transfer, can greatly facilitate their wider use, improving patient care and serosurveillance in the community.

ACKNOWLEDGMENT

The authors gratefully acknowledge the contributions of Dr. SR Swarna at Jawaharlal Institute of Medical Education and Research, Karaikal, Puducherry, India to writing the chapter.

FURTHER READINGS

Balachandra D, Rahumatullah A, Lim TS, Mustafa FH, Ahmad H, Anuar NS, Noordin R. A new antigen detection ELISA for the diagnosis of *Strongyloides* infection. *Acta Tropica.* 2021;221:105986.

Chappell CL, Wright JA, Coletta M, Newsome AL. Standardized method of measuring acanthamoeba antibodies in sera from healthy human subjects. *Clin Diagn Lab Immunol.* 2001;8:724–730.

Chaya DR, Parija SC. Evaluation of a newly designed sandwich enzyme linked immunosorbent assay for the detection of hydatid antigen in serum, urine, and cyst fluid for diagnosis of cystic echinococcosis. *Trop Prasitol.* 2013;13:125–131.

Crowther JR. *Methods in Molecular Biology: The ELISA Guidebook.* New York: Humana Press; 2009a.

Hosseini S, Vázquez-Villegas P, Rito-Palomares M,Martinez-Chapa SO. *Enzyme Linked Immunoabsorbent Assay (ELISA): From A to Z.* Singapore: Springer Nature; 2018.

Parija SC, Ananthakrishnan N. Evaluation of stabilized cells in the indirect haemagglutination test for echinococcosis. *J Med Microbiol.* 1985b;19:95–98.

Parija SC, Ravinder PT, Rao RS. Detection of hydatid antigen in urine by counter-current immunoelectrophoresis. *J Clin Microbiol.* 1997;35:1571–1574.

Ravinder PT, Parija SC, Rao RS. Urinary hydatid antigen by co-agglutination, a cost effective and rapid test for the diagnosis ofcystic echinococcosis in a rural or field settings. *J Clin Microbiol.* 2000b;38:2972–2974.

Swarna SR, Parija SC. Evaluation of Dot-ELISA and enzyme-linked immune-electrotransfer blot assay for detection of a urinary hydatid antigen in the diagnosis of cystic echinococcosis. *Trop Prasitol.* 2012;12:38–44.

Vogt Jr RF, Phillips DL, Henderson LO, Whitfield W, Spierto FW. Quantitative differences among various proteins as blocking agents for ELISA microtiter plates. *J Immunol Methods.* 1987;101:43–50.

11 Fluorescent Dye-Based Tests

INTRODUCTION

The use of fluorescent dyes as a coupling agent began with the seminal work of Coons and associates in 1942. They were the first to demonstrate that antibodies can be conjugated with fluorescein isocyanate, and they used the conjugate to demonstrate pneumococcal antigen in tissues. Although the discovery did not receive much attention initially, interest in the technique and its application saw an upswing in the 1950s, and before the advent of enzyme immunoassays, much work was done in immunofluorescence assays. Even in the 21st century it occupies an important place in the serodiagnosis for detecting both antigen and antibody in bodily fluids and tissue specimens.

PRINCIPLE OF IMMUNOFLUORESCENCE

A fluorescent dye absorbs light in the blue-green to ultraviolet range, which causes excitation of the electron due to absorption of energy from the photon. When the molecules return to the resting stage, photons are emitted at a different and longer wavelength, which can be visualized. The visualization of the fluorescence is done using a microscope with dark field condenser and suitable filters to create a dark background against which the fluorescence can be seen (Figure 11.1).

Various conventional light sources like mercury, xenon, and halogen lamps are used to generate the ultraviolet light. In recent years, bright, single-wavelength light-emitting diodes (LEDs) are increasingly being used. LED has many advantages over the conventional light sources. They offer higher luminous efficiency, compact size, lower power consumption, longer operational lifetimes (10,000–15,000 hours), and cost effectiveness.

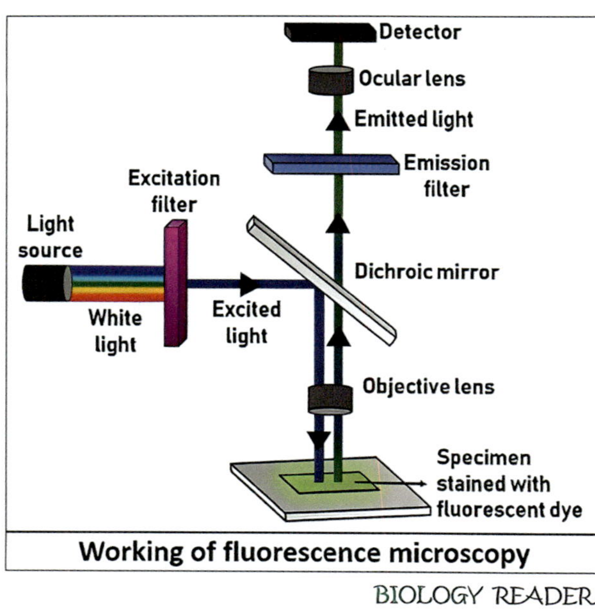

FIGURE 11.1 Principle of fluorescence microscope.

(Picture credit: https://biologyreader.com/fluorescence-microscopy.html.)

DOI: 10.1201/9781003644224-14

For the immunofluorescent test formats, the antibody is labeled with a fluorochrome probe, much like the labeling of an enzyme. The fluorochromes that are commonly used include fluorescein isothiocyanate (FITC, the most common), TRITC (Tetramethylrhodamine-5-(and 6)-isothiocyanate) used in conjunction with FITC, and cyanins (Alexa Fluor® dyes). Other dyes used infrequently are phycoerythrin, europium (β-naphthyl trifluoro acetone), and lucifer yellow VS.

Fluorescein shows maximum absorption at 490 to 495 nm and emits a green light at 517 to 520 nm, while TRITC absorbs at 550 nm and emits red color at 580 to 585 nm. Since their absorbance and emission patterns differ, they can be used together. The antibody may be a primary or a secondary antibody raised in animals. The working principle is similar to enzyme-based assays except that after the capture step, the visualization of the antigen–antibody complex is done using the fluorescent microscope when bright fluorescence of the complex may be observed.

TYPES OF IMMUNOFLUORESCENCES

The two formats commonly used are the direct and indirect immunofluorescence assays.

Direct Immunofluorescence Assay

The direct immunofluorescence assay (DFA)technique is used for detection of an antigen. The patient's sample containing the putative antigen is layered on a clean glass slide, and fluorescein-tagged monoclonal or polyclonal antibody is added and observed under the fluorescent microscope. This can also be used for tissue sections fixed on the slide. DFA are commercially available for *Giardia* and *Cryptosporidium*.

Indirect Immunofluorescence Assay

Indirect immunofluorescence assay (IFA) is the most popular format, used for either antigen or antibody detection. It obviates the need for using separate labeled antibody samples as required in DFA. The antigen or antibody is already coated on the slide to which the patient's sample is added. If the corresponding analyte is present in the sample, it will bind to the corresponding molecules on the slide. The antigen–antibody complex is then detected with a fluorescein-labeled secondary antibody (anti-human antibody) raised in rabbit, mouse, or goat.

Both DFA and IFA techniques have also been studied for *Trypanosoma brucei*, *Trypanosoma cruzi*, *Babesia* spp., malaria parasite, *Toxoplasma gondii*, *Leishmania* spp. (Figure 11.2), and *Strongyloides stercoralis*.

The following sections outline the in-house development and design of the antibody labeling with a fluorescein isothiocyanate for detection of parasite antigen and the indirect fluorescent antibody test for schistosomiasis for demonstration of antibodies.

Antibody Labeling with a Fluorescein Isothiocyanate for Detection of Parasite Antigen

Fluorescein isothiocyanate (FITC) has two isomers, I and II. FITCI has the isothiocyanate form at C_4 position on the benzene ring while FITC II has it at the C_5 position. Since FITC I is easily produced in pure form and less expensive, it is commonly used for labeling of antibodies. The isothiocyanate group reacts with amino terminal and primary amines in protein molecules.

Reagents and Equipment Required

Reagents: PBS (pH 7.4); sodium bicarbonate buffer (pH 9.0); fluorescein isothiocyanate; dimethyl sulfoxide; ammonium chloride; xylene cyanol; glycerol; bovine serum albumin; sodium azide.

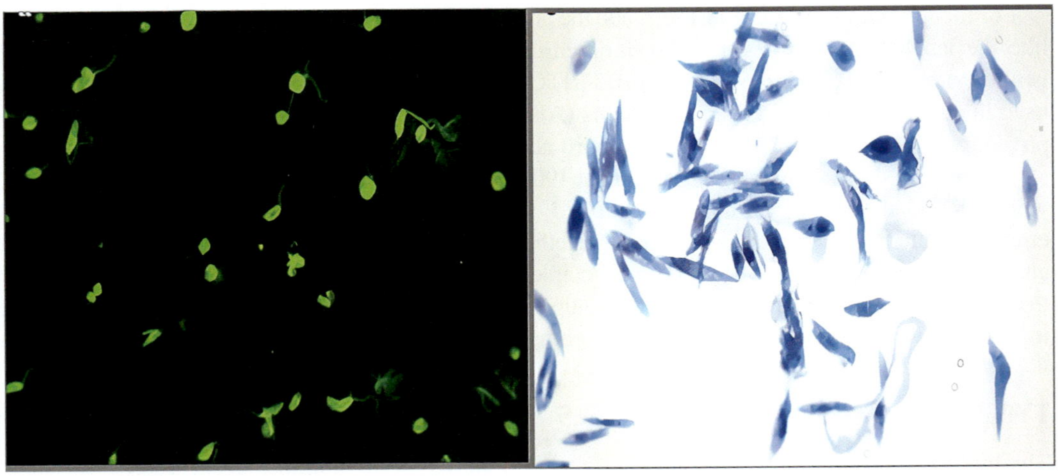

FIGURE 11.2 Left: IFAT of promastigotes of *Leishmania infantum* stained by FITC. Right: Cultured promastigotes of *L. infantum*.

(Photo adapted from Mendoza-Roldan JA, Latrofa MS, Iatta R, Manoj R, Panarese R, Annoscia G, et al. Detection of *Leishmania tarentolae* in lizards, sand flies, and dogs in southern Italy, where *Leishmania infantum* is endemic: hindrances and opportunities. *Parasit Vectors*. 2021;14(1):461.) (This article is licensed under a Creative Commons Attribution 4.0 International License, which permits use, sharing, adaptation, distribution, and reproduction in any medium or format, if you give appropriate credit to the original author(s) and the source. http://creativecommons.org/licenses/by/4.0/. The Creative Commons Public Domain Dedication waiver http://creativecommons.org/publicdomain/zero/1.0/.)

Equipment and labware: dialyzer; gel filtration column; spectrophotometer; quartz cuvettes; micropipettes; stirrer.

Others: any polyclonal or monoclonal antibody of interest.

Preparation of the Conjugate

Antibody Processing

Any type of antibody, either purified polyclonal antibody from hyperimmune serum or monoclonal antibody, can be used depending on the assay and availability.

- For DFA, these antibodies conjugated to FITC can be used to detect the antigen.
- For IFAT, either goat or rabbit anti-human immunoglobulin can be used as a secondary antibody. Such tagged antibodies are also available commercially.
- The antibody to be conjugated should be free of contaminating proteins. To remove other amines, the antibody solution should be dialyzed against PBS (pH 7.4) overnight at 0°C–5°C.
- A solution of the antibody is prepared in a 0.1 M sodium carbonate buffer (pH 9.0) at a concentration of 2 mg/mL or more. The buffer should be freshly prepared.

FITC Labeling

FITC is dissolved in anhydrous DMSO at 1 mg/mL immediately before labeling.

- 50 µL of FITC is added very slowly per 1 mL of antibody solution in 5-µL aliquots with continuous gentle stirring.
- Once the total amount has been added and mixed, the mixture is kept at 4°C for 8 hours in the dark.
- NH_4Cl is added to a final concentration of 50 mM and incubated for 2 hours at 4°C.

- Xylene cyanol to 0.1% and glycerol to 5% of the total volume are added.
- The excess unbound FITC is removed by gel filtration using a gel matrix having an exclusion limit of 20,000 to 50,000. Before starting the column flow, the solution is carefully layered over the column. Then the column is opened and the mixture is allowed to flow down the column. When all of it reaches the column bed, PBS is added to the top of the column and connected to a buffer supply.
- Two bands can be seen in the column. The faster-moving band is the conjugated protein, which elutes first and is visible at room illumination, while the slower-moving band is the free unbound FITC with xylene cyanol, which elutes after the washing with PBS.

Determination of FITC/Protein Molar Ratio (F/P)

- The F/P molar ratio is the ratio of moles of FITC to moles of protein in the conjugate. To calculate it, the absorbance of the conjugate sample at 280 nm and then at 495 nm must be measured.
- The conjugate sample is kept in a quartz cuvette, and readings are taken at the two wavelengths. The absorbance at 280 nm should be 0.2–1.4.
- If the reading at 280 nm is outside this range, suitable dilution of the mixture should be performed.
- F/P ratio is calculated as:

$$\text{Molar F/P} = 2.77 \text{ x } A_{495}/A_{280} - (0.35 \text{ x } A_{495})$$

- The protein concentration of the fluorescein-IgG conjugate is calculated using the following formula:

$$\text{IgG (mg/mL)} = A_{280} - (0.35 \text{ x } A_{495})/1.4$$

where 1.4 is the A_{280} of IgG for most species at a concentration of 1.0 mg/mL at pH 7.0, while 0.35 is the correction factor.

- The F/P ratio should be between 0.3 and 1.0. Lower ratios will yield weak signals while higher ratios will give high background noise.

Storage

For storage of the conjugate after determination of the F/P molar ratio, 1% (w/v) BSA and 0.1% (w/v) sodium azide are added to the conjugate and stored at 2°C−8°C protected from light.

For more details of the procedure and methods to label other antibody classes like IgM or IgA, please consult the Sigma Adrich catalog.

Designing the Indirect Fluorescent Antibody Test for Schistosomiasis by Demonstration of Antibodies

Schistosomiasis is a disease confined to a few endemic countries but is being encountered in increasing numbers in non-endemic countries due to immigration and tourism. Because of low burden of the parasite, making a microscopic diagnosis in these patients is difficult, and serological tests to detect antibodies in such individuals are invaluable.

The following IFAT procedure may be used for a quick diagnosis.

Reagents and Equipment Required

Reagents: formalin solution; PBS; ethanol; ethyl ether; p-phenylenediamine; glycerol; normal saline.

Equipment and labware: freezing-microtome; high-speed homogenizer; vacuum dryer; centrifuge; multi-well glass slides; syringe and needle (without bevel); epifluorescent microscope; micropipettes.

Others: *Schistosoma mansoni* adult worm; FITC-tagged rabbit anti-human immunoglobulin.

Development of the Test

Antigen preparation: Two methods may be adopted depending on the laboratory capacity.

(i) *Method 1*

S. mansoni adult schistosomules are obtained from outbred CF1 mice that had been infected with the cercariae 8 weeks earlier.

- The parasites are kept in 2% formalin for fixation and preservation.
- Using a freezing-microtome, sections of the parasites are made of 5μm thickness, and individual sections are placed on multi-well slides and stored in PBS at −20°C.

(ii) *Method 2*

S. mansoni adult worms are harvested from liver and portal system of mice and are carefully washed in saline and in distilled water.

- The worms are vacuum dried at −20°C.
- Dried worms are triturated for 5 minutes at 45,000 rpm in a Virtis 45 homogenizer in cold absolute ethanol in an ice salt bath.
- The sediment is then obtained by slow centrifugation at 3,000 rpm at 4°C.
- The sediment is re-suspended in cold anhydrous ethyl ether, shaken, and re-centrifuged.
- The final product is vacuum dried and may be kept for extended periods at −20°C without any deterioration of its antigenic property.
- Prior to use, the parasite particle is suspended in PBS and carefully layered on multi-well slides using a hypodermic needle without bevel to make a thin film.
- The material is then fixed on the slides by drying at 37°C for 30 minutes.

Test Procedure

- The slide with the antigen preparation is taken and 20 μL patient serum (Undiluted, and double diluted 1:10 to 1:160 in PBS) is carefully layered over it.
- The slides are incubated at room temperature for 30 min in a wet chamber.
- The preparation is then washed thrice with PBS to remove unbound antibodies.
- This is followed by addition of 20 μL FITC tagged rabbit anti-human immunoglobulin, either obtained commercially (various suppliers like Sigma-Aldrich, Fisher Scientifics, Agilant, etc.) or prepared as mentioned above.
- The preparation is incubated again for 30 minutes at room temperature in a wet chamber, followed by three times washing with PBS.
- It is then dried and mounted in glycerin (pH 8.0) with a coverslip and observed under the fluorescent microscope with a 495-nm excitation filter.
- Adding p-phenylenediamine to the mounting helps prevent fading of the fluorescence

Results and interpretation: An intense green fluorescence of worm particles can be observed if homogenized preparation of the parasite has been used.

If sections of the parasite have been used, two types of fluorescence are seen. One type is a clear fluorescence at the outer surface of the worm (membrane-bound antigen), and the other type is on the internal gut surface of the parasite (gut-associated antigen).

It is best to use epifluorescence where the incident light first passes through the objective lens (acting as condenser) and then focuses on the specimen.

Quality assurance: Appropriate positive control and negative controls must be included in each test. Secondary antibody control (no primary antibody added) can be added to check that the secondary antibody is not producing false positives. Autofluorescence controls are sometimes used (no primary or secondary antibody) to check whether the antigen is producing autofluorescence.

TABLE 11.1
Available Commercial Immunofluorescent Kits for Parasitic Infections

	Antigen Detection	Antibody Detection
Protozoa	*Giardia intestinalis* (stool) *Cryptosporidium* spp. (stool)	*Toxoplasma gondii* *Leishmania donovani*
Trematode	—	*Schistosoma mansoni*.

Note: For the suppliers of these kits, please refer to Appendix C.

Troubleshooting

High background fluorescence: This occurs due to drying of the slide at some stage of preparation. Hence, incubation should always be carried out in a wet chamber.

Nonspecific fluorescence: This occurs due to cross-reactivity between the conjugated (secondary) antibody and cellular protein antigens. It is identified by using secondary antibody control where only the secondary antibody is applied to the preparation and primary antibody (patient's serum) is omitted. To prevent this, the dilution factor of the conjugate can be increased or blocking agents such as 1% bovine serum albumin or Tween-20 can be added to the conjugate diluent. Other methods to reduce nonspecific staining include adsorption of antibodies with human liver tissue homogenates or removal of super-conjugated antibodies by ion-exchange chromatography.

Commercially Available Immunofluorescent Kits

Very few DFA or IFAT are available commercially for parasitic diagnosis (see Table 11.1). However, individual parasite antigens as well as fluorophore-labeled secondary antibodies are available from various suppliers, the full list of which can be found in Appendix C.

CONCLUSION

Fluorescent antibody tests for parasites enjoyed their heyday in the 1950s and 1960s, prior to the advent of enzyme-based assays. With the introduction of ELISA, fluorescent dye–based tests have seen a decline for reasons such as subjective interpretation, lack of automation, lack of commercial suppliers of kits, and inability to test many samples simultaneously. However, they remain an attractive option for developing in-house diagnostic tests because of the ease of design and commercial availability of fluorescence-tagged secondary antibody.

FURTHER READINGS

Camargo ME, Hoshino S, da Silva LC. A slide fluorescentantibody technique with adult worm antigen for the serological diagnosis of *Schistosomiasis mansoni*. *Rev InstMed Trop São Paulo.* 1965;7:327–331.

Coons AH, Creech HJ, Jones N, Berliner E. The demonstration of pneumococcal antigen in tissues by the use of fluorescent antibody. *J Immunol.* 1942;45:159–170.

Product Information. FluoroTag™ FITC Conjugation kit, Sigma-Aldrich Co. LLC, 2021. Available at: https://www.sigmaaldrich.com/Sigma/Bulletin. Accessed March 19, 2025.

Sherwin T, Read M. Immunofluorescence of parasites. In Hyde JE, editor, *Methods in molecular biology, Vol 21. Protocols in Molecular Parasitology*. Totowa, NJ: Humana Press; 1993

Tarp B, Black FT, Petersen E. The immunofluorescence antibody test (IFAT) for the diagnosis of schistosomiasis used in a non-endemic area. *Trop Med Int Health.* 2000;5:185–191.

12 Immunoblot Tests

INTRODUCTION

Immunoblotting technique, popularly known as Western blot, is used to identify and detect specific protein molecules in a mixture of proteins extracted from tissues. In infectious disease diagnostics, it is used as a sensitive method to detect multiple antibodies produced in the body in response to a number of antigens. The technique was developed simultaneously in 1979 by George Stark and his team in the United States and by Harry Towbin's group in Switzerland.

The technique consists of three steps: separation of protein molecules by size using electrophoresis; transfer of the separated proteins to a solid support; and detection and visualization of these proteins by primary and secondary labeled antibodies. At one time it was extensively used for the laboratory confirmation of HIV infection. The technique continues to be a powerful and commonly used method not only in immunological diagnostics but also in identifying and quantifying protein expression in the field of proteomics.

PRINCIPLE OF IMMUNOBLOTTING TECHNIQUE

Separation of Antigens

An organism contains many different types of proteins that are antigenic. The first step is to separate them based on molecular weights. The organism is first disrupted using chemical agents, followed by electrophoresis in sodium dodecyl sulphate–polyacrylamide gel (SDS-PAGE). In the presence of SDS, the negatively charged proteins move in the electrophoresis gel according to their molecular weights, the lighter ones being situated farthest from the well.

Transfer of Antigens

In this step, the bands of proteins generated by SDS-PAGE are transferred (blotted) to nitrocellulose membranes or other materials. The transfer can be done in two ways: capillary transfer or using an electric field (electroblot). In this procedure, the antigen bands get aligned on the blotting material.

Detection of Antibodies

The patient's serum is added to the membrane, followed by the enzyme conjugate and the substrate-chromogen similar to ELISA. Colored bands are produced if corresponding antibodies to the blotted antigens are present. A Western blot band profile is thus created and interpreted.

TYPES OF IMMUNOBLOT

These include Western blot, enzyme-linked immunoelectrotransferblot (EITB), line immunoassay (LIA), and recombinant immunoblot assay (RIBA).

Western Blot

This is the classical type in which proteins are blotted on the membrane using capillary action. In this procedure, a dry absorbent material is placed on top of the gel, and the blotting membrane is placed between the gel and the absorbent material. The gel is then placed on a filter paper soaked in

DOI: 10.1201/9781003644224-15

a suitable buffer solution. By capillary action, the buffer is drawn through the gel carrying the protein bands, which then bind to the blotting material. This is an easy method not requiring expensive instrumentation needed for protein blotting as in electroblot. It is useful for blotting isoelectric focusing gels with zero overall charged proteins that are located at the isoelectric points and hence difficult to transfer by electrophoresis. However, in this method, there is incomplete transfer of the proteins from the gel, and it is a time-consuming process requiring 12–18 hours of incubation.

Enzyme-linked Immunoelectrotransfer Blot (EITB)

In this format, the proteins are transferred from the electrophoresis gel to the membrane using an electric current applied perpendicular to the gel surface, with the positive electrode at the membrane end. The gel with the membrane is sandwiched between sponge or fiber pads and filter papers to protect the gel and the membrane. The negatively charged proteins migrate from the gel and get absorbed by the membrane. Although the whole setup can be devised manually, it is more easily done in commercially available electroblot instruments. It is much faster (45–90 minutes, depending on gel thickness) compared with capillary transfer and much more efficient, with complete transfer of the proteins to the blotting membrane.

Line Immunoassay (LIA)

The format of LIA is the same except that instead of blotting, synthetic or recombinant antigen fractions are painted/smeared at different locations on a nitrocellulose membrane. It has several advantages over the classical WBT, including clearer background and use of standardized and optimum concentration of the antigen. Suitable controls can be applied in the same strip, eliminating the problem of indeterminate results by getting objective results using automatic readers.

Recombinant Immunoblot Assay (RIBA)

This method is similar to LIA except that only recombinant antigens are used.

IMPORTANT CONSIDERATIONS IN DESIGNING AN IN-HOUSE IMMUNOBLOT SYSTEM

Antigen Preparation

Cells or tissues of parasites are commonly used to extract the antigens used for electrophoresis. Sonication and homogenization techniques are used for the purpose. All the steps should preferably be carried out in a cold condition using a lysis buffer with protease inhibitors to prevent protein denaturation. Homogenization can be achieved using ultrasonic, mechanical, or manual methods. Following extraction, the protein concentration should always be checked spectrophotometrically. This will help in choosing the appropriate volume that should be loaded in electrophoresis chamber.

The antigen is then diluted with the loading buffer containing glycerol and a tracking dye. Glycerol helps sink the antigen in the electrophoresis well while the tracking dye (Bromophenol blue) helps monitor the separation of the antigen. A preliminary step of heating the antigen before diluting ensures that higher-order structures are denatured but the sulfide bridge remains intact.

Electrophoresis

Polyacrylamide gel with sodium dodecyl sulphate (SDS-PAGE) is used to separate the antigenic fractions. SDS is an ionic detergent and binds strongly to the protein molecules (one SDS ion for every two amino acids) with a net negative charge of the complex. It causes proteins to acquire an approximately constant charge-to-mass ratio, allowing them to migrate through the gel at a rate proportional to their molecular mass. The gel slab used is a gradient gel.

The higher (stacking) gel contains lower concentration of polyacrylamide (larger pores) and is slightly acidic (pH 6.8). The lower (resolving/separating) gel has higher concentration (smaller pores) and higher pH of 8.8 (alkaline). Care should be taken while preparing the gels. The stacking gel is prepared in a lower concentration of the buffer solution (e.g., 0.125 mol/L of Tris-HCl) compared with the resolving gel (0.375mol/L of Tris-HCl).

The stacking gel cannot separate the proteins efficiently but helps them form sharp, well-defined bands. The resolving band helps in better separation of the proteins, since smaller fractions can move rapidly compared with larger-mass protein fractions. Gels are prepared by pouring them between two glass or plastic plates. The proteins with a negative charge will move toward the positive electrode (anode).

A molecular ladder of known molecular weights (obtained commercially) is included during the run to characterize and label the separated antigen according to their molecular weights. For accurate determination of molecular weights, a standard curve needs to be prepared using relative mobility (migration distance of dye front/migration distance of the protein) in x-axis and the log of the molecular mass (y-axis), which will be a straight line. The mass of an unknown is accurately determined from the linear curve.

Buffer Solutions

Buffer solutions for the immunoblot include the following:

a. *Lysis/antigen preparation buffer*
Tris-Triton or Tris-HCl (20 mM Tris-HCl, pH 7.5) buffers with protease/phosphatase inhibitors are commonly used for parasites. Tris-Triton buffer consists of 50 mM Tris (pH 8.0), 150 mM NaCl, and 1% NP-40 or Triton X-100. Triton X-100/ NP-40 are nonionic detergents that help solubilize the proteins without denaturing them.

b. *Loading/sample buffer*
The sample to be electrophoresed is mixed with the loading buffer. It contains Tris-HCl to maintain pH at 8.3; sodium dodecyl sulphate (SDS) for protein denaturation and adding net negative charge to proteins; dithiothreitol to disrupt disulfide bonds and help in denaturation; glycerol to allow the sample to sink in the well by increasing the total density; and bromophenol blue to allow the tracking of low-molecular-weight protein fraction.

c. *Running buffer*
Tris-glycine is the most used buffer for running the gel. Glycinate ions form trailing end in the stacking gel and move faster than the protein ions. Other running buffers that can be used are Tris-HCl, acetate buffer (useful for separation of high-molecular-weight fractions, 100–500 kDa in size), MES, and MOPS, among others.

d. *Transfer buffer*
The Towbin system buffer is most commonly used. It has the following composition:

- Tris-HCl: 25 mM; maintains the pH at 8.3.
- Glycine: 192 mM; prevents the gel from swelling.
- Methanol: 20% [v/v]; prevents the gel from swelling, removes SDS from proteins that help in better binding to the membranes.
- SDS: 0.01%; increases protein transfer efficiency.
- For better transfer of high-molecular-weight proteins (>150kDa), cyclohexylamino-l-propanesulfonic acid (CAPS) can replace glycine, particularly in semi-dry transfers.

Transfer Membrane

Although various types of membranes can be used for blotting, the two most used are nitrocellulose and polyvinylidene difluoride (PVDF).

a. *Nitrocellulose membrane*
   It has several advantages, such as high protein-binding capacity, to distinguish between both high- and low-molecular-weight proteins, to immobilize proteins, and to provide a platform not only for chromogenic detection but also for chemiluminescence and fluorescence markers. However, it is brittle and cannot be used for re-probing.
b. *PVDF*
   It gives a much better mechanical support and is not brittle, hence can be reused for probing with different antibodies subsequent to the initial run. PVDF needs to be activated with methanol for 15 seconds before use. One disadvantage of PVDF is that it gives higher background staining, which needs careful washing.

   Evidence of transfer: Before proceeding with the blocking step and running the assay, it is a good practice to verify that satisfactory transfer has occurred to save time and resources. It can be done using a stain or by stain-free method. Ponceau S is a diazo compound that is negatively charged if used with acetic acid; it binds to the positively charged amino acids' nonpolar protein regions. The staining reagent is prepared by mixing 0.01% [w/v] Ponceau S in 1% acetic acid [v/v], and the membrane is stained for 1minute, when the protein bands are stained red or pink. After visualization, the stain can be easily removed by washing the membrane with distilled water.

   The staining procedure is also useful for normalization of the Western blot to identify technical errors such as unequal sample loading in electrophoresis and has been found to be superior to the β-actin housekeeping protein.

   The stain-free method utilizes the presence of tryptophan in the transferred proteins, which undergoes modification and fluoresces strongly when exposed to ultraviolet light. A handheld transilluminator can be used to detect the transferred bands.

Blocking Material

Blocking materials help prevent nonspecific binding of antibodies to the membrane and reduce background noise as well as false-positive results.

Nonfat (skim) dried milk diluted in Tris-buffered saline Tween-20 (TBST) is the most commonly used blocking agent, since it is most economical. Alternately, 5% bovine serum albumin (BSA) can be used. The proteins present in milk or BSA occupies the areas where the transferred proteins are absent. Tween-80 prevents any weak binding of milk proteins or BSA to the transferred protein bands. It should be noted that milk proteins contain biotin and phosphoproteins, which are incompatible with biotin- or alkaline phosphatase–based assays. To address the issue, it is a good practice to block the membrane with the skim milk first. Subsequently, the sample containing the primary antibody is incubated with 5% BSA overnight at 4°C in a shaker. The incubated antibody can be reused if the assay gives ambiguous results.

This section describes the in-house design and optimization of the EITB protocol for cysticercosis.

Design and Optimization of EITB Protocol for Cysticercosis Including Neurocysticercosis

The diagnosis of cysticercosis including neurocysticercosis depends both on radiological imaging (CT and preferably MRI) and serological testing by ELISA or EITB.

As per CDC, EITB is a better technique in terms of sensitivity and specificity for antibody detection compared with ELISA. EITB positivity can be detected as early as 5weeks after exposure and demonstrates a sensitivity of 98% and a specificity of 100% in patients harboring multiple cysts; however, it is less sensitive in patients with a single cyst. It is reported to be unaffected by any immunosuppressive conditions in the patients. Here a generic protocol is described where, except for the antigen preparation for the assay, other steps are essentially similar to the procedures described by various authors and the protocols listed in the Further Readings section.

Reagents and Equipment Required

Reagents: phenylmethylsulfonyl fluoride (PMSF); protein-free hybridoma medium (PFHM); antibiotic supplement; PBS; Tris base; glycine; methanol; PBS-Tween buffer; nonfat dry milk; Tris–HCl; acrylamide bisacrylamide; SDS; dithiothreitol; ammonium persulphate; tetramethyl ethylene diamide (TEMED); Tris-glycine; NaCl; KCl; Tween-20; sodium azide; 3,3′-diaminobenzidine (DAB) or tetramethyl benzidine (TMB); Ponceau S or stain-free stain.

Equipment and Labware: ultra-centrifuge; tissue culture flasks; centrifuge with a compatible rotor for ultrafiltration; ultrafiltration centrifuge tubes; YM-10 ultrafiltration membrane; hot plate; electrophoresis apparatus with a power bank; glass plate; filter papers; sponge pieces; PVDF/nitrocellulose membrane; micropipettes.

Others: hydatid cyst fluid; *Taenia crassiceps*; BALB/c mice; horseradish peroxidase (HRP)-conjugated goat antibody.

Design and Optimization of the Test

Antigen Preparation

Different workers have used various cysticeral antigens, which are briefly discussed here. Interested readers can refer to the original studies, listed in the Further Readings section, for the details.

Cyst fluid antigen: Infected pigs from a slaughterhouse are selected and the cysts from muscle tissues are isolated.

- The aspirated cyst fluid is subjected to sonication followed by ultra-centrifugation.
- The supernatant is separated and treated with phenylmethylsulphonyl fluoride (PMSF, a protease inhibitor).
- After determining the protein content using the Bradford assay, the prepared antigen can be stored at −80°C.

Taenia crassiceps excretory/secretory (ES) antigen: *T. crassiceps* is a close relative of *T. solium* first isolated in 1962. It can be easily maintained in mice by serial intraperitoneal passage of the cysts, making it a useful model for *in vitro* and *in vivo* studies.

- BALB/c mice are inoculated with 5–10 cysts- and about 900 cysts could be harvested after 90 days.
- The cysts are washed thrice in protein free hybridoma medium (PFHM).
- Twenty tissue culture flasks each containing 10 mL of PFHM and antibiotic supplement (100IU penicillin, 50 μg streptomycin, and 2.5 μg amphotericin B/mL) are inoculated with approximately 45 cysts and incubated at room temperature for 144 hours.
- The flask is examined daily for appearance of the parasites and vesicle budding.
- At 24, 48, and 96 hours, the medium is replaced. The old culture medium at each change and the one at the end of incubation are collected and pooled with addition of required amounts of PMSF.
- The pooled materials are concentrated by centrifugation using ultrafiltration with YM-10 membrane and the antigen preparations designated ES24, ES48, ES96, and ES144 were stored at −20°C.

- To test the viability of the cysts, four mice are inoculated intraperitoneally with 15 cysticerci each, sacrificed after 90 days, and examined for the cysts in the peritoneal cavity.

LL-GP T. solium cysticercus antigen: The lentil-lectin glycoprotein antigens (LL-GP) obtained from *T. solium* cysts are considered the preferred antigens and the gold standard for an EITB assay. Developed by Tsang and coworkers at the Centers for Disease Control and Prevention (CDC, USA) and first reported in 1989, it is a set of seven purified antigens, namely GP 50, 42–39, 24, 21, 18, 14, and 13 kDa.

Its preparation is technically demanding and time consuming. However, LDBIO Diagnostics (France) is one commercial source of the Western blot kit with the seven antigens and with distributors worldwide, including India. Each kit comes with sensitized and precut strips, the ready-to-use liquid reagents, and the positive control serum.

Antigen Quantification

Whatever the source or technique of preparation, the volume of protein extract to be loaded in the wells is determined using the formula: Concentration = mass/volume, to ensure loading of 50 μg of antigen in each well.

- To the protein antigen, 5 μL of the sample buffer and the necessary amount of double-distilled water are added to make a final volume of 15 μL to be dispensed in each well.
- Sample or loading buffer is prepared as follows:
  - Tris-HCl (1M): 2 mL
  - Dithiothreitol (1M): 4 mL
  - SDS: 0.8 g
  - Bromophenol blue: 40mg
  - Glycerol: 3.2 mL

Final volume made up to 10 mL by adding double-distilled water.

Gel Preparation

Stacking gel (10%) is prepared with the following ingredients:

Double-distilled water: 3 mL
1 M Tris–HCl: 2.1 mL (pH 8.9)
30% acrylamide bisacrylamide: 2.8 mL
10% SDS: 80 μL
10% ammonium persulphate (APS): 56 μL
tetramethyl ethylene diamide (TEMED): 6 μL

- The rack is assembled and the stacking gel solution is carefully added with a suction pipette to the bar holding the glass plate.
- Water is added to the top, and the gel is allowed to solidify (15–30 minutes).
- The separating gel (6%) is prepared with the following ingredients:
  Double-distilled water: 2 mL
  1M Tris–HCl: 400 μL (pH 6.7)
  30% acrylamide bisacrylamide: 600 μL
  10% SDS: 36 μL
  10% AP: 24 μL
  TEMED: 4 μL

- The water is removed and the resolving gel solution is carefully added to overlay the stacking gel by slightly tilting the rack.
- Once the solution solidifies, the comb can be inserted.

  Note: Both gels should be prepared at the same time, or else the APS and TEMED should be added at the end.

Gel Electrophoresis

- The electrophoresis chamber is filled with the running buffer (Tris-glycine).
- The gel is placed in the chamber, and the power supply is connected.
- The comb is gently removed, and the wells are filled with the sample (15 µL) with tracking dye (bromophenol blue, 6 µL).
- The gel is run at 60V, and when all the samples have entered the stacking gel, the voltage is increased to 140V until the dye front reaches the bottom of the gel. Alternately, a constant voltage of 100V can be applied.

Transfer to Membrane

Either the tank blotting (wet transfer) system, which is commonly used and takes about 60–90 minutes for transfer, or the rapid semi-dry method, which takes 3–10 minutes, can be used. The blotting apparatus for both types of transfer is commercially available. Here we describe the tank or wet system of electrotransfer.

- Six filter papers, two sponge pieces, and one PVDF/nitrocellulose membrane are cut to the same size as the gel slab.
- The PVDF is activated by methanol for 15 seconds, and the sponge and filter paper pieces are moistened well with the transfer or Towbin buffer, which can be prepared as follows:
  - Tris base (25 mM): 6.04 g
  - Glycine (192 mM): 28.8 g
  - Methanol (10% [v/v]): 200 mL
  - SDS: 0.01%
  - For better transfer of high-molecular-weight proteins (>150kDa), cyclohexylamino-L-propanesulfonic acid (CAPS) can replace glycine, particularly in semi-dry transfers.

The above ingredients are mixed in 1,600 mL of double-distilled water and the final volume is made up to 2,000 mL.

- The stack is made as follows: sponge-filter papers (3)–gel-membrane-filter papers(3)–sponge (Figure 12.1); the sandwiched material is transferred to the blotting apparatus kept on ice to maintain a temperature of −4°C.
- Electrophoresis is carried out with the blotting membrane toward the positive electrode. A constant voltage of 60 V is applied. Typically, it takes 90 minutes for the transfer to complete.
- The transfer can also be done by the modern semi-dry or the dry systems, and the equipment is available from Bio-Rad and Thermo Fisher Scientifics. In the semi-dry system, transfer can occur in 5–10 minutes, and in as little as 3 minutes in the dry method. The manufacturer's instructions should be followed when using these instruments.

Blocking

Following transfer, the membrane is carefully removed and gently wiped with dry gauze to remove any gel material.

- With a lead pencil, lines are softly drawn across the top and bottom of the membrane, which helps alignment if the membrane needs to be cut into strips. Ponceau S or Stain Free can help in visualizing the transferred bands and cutting the strips.

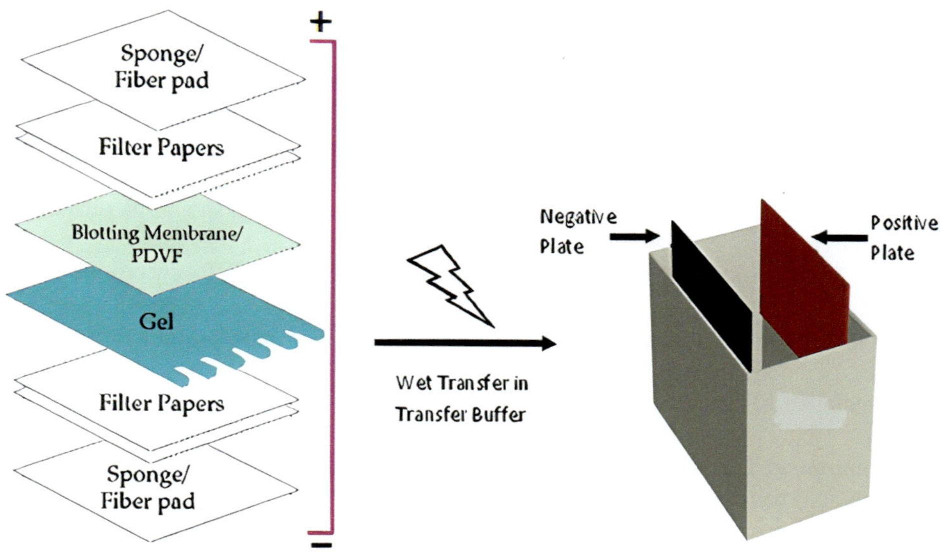

FIGURE 12.1 Assembly of transfer system.

(From Mahmood T, Yang P. Western blot: technique, theory, and trouble shooting. ***North Am J Med Sci.*** **2012;4:429–434.) (This is an open-access article distributed under the terms of the Creative Commons Attribution-Non-Commercial-Share Alike 3.0 Unported, which permits unrestricted use, distribution, and reproduction in any medium, provided the original work is properly cited.)**

- The membrane is then blocked with 5% skim milk in TBST for 1 hour.
  TBST is prepared as follows:
  - ➢ Distilled water: 800 mL
  - ➢ NaCl: 8.8 g
  - ➢ KCl: 0.2 g
  - ➢ Tris base: 3 g
  - ➢ Tween-20: 500 μL

  pH is adjusted to 7.4, and distilled water is added to take the final volume to 1,000 mL. It is sterilized by autoclaving/filtration and stored.

Storage of Strips

It can be done for up to 6 months if kept at −60°C covered with a filter paper saturated with PBS. Alternatively, the strips may be kept at −4°C covered with PBS-saturated filter paper with added 0.1% sodium azide.

Test Procedure

This essentially follows the ELISA procedure. The membrane strips are laid on plastic trays, and the following sequential steps are performed:

- Serum or CSF of patients is tested neat and with 1:50 and 1:10 dilutions, respectively. Dilution is done in PBS/Tween buffer with 5% nonfat dried milk.
- 0.5 mL of diluted serum/CSF is laid on the plastic trough or specially made incubation tray, and the strip is immersed in it with blotted surface facing up. It is incubated for 1 hour.
- Unbound serum/CSF components are removed by gentle washing with warm (56°C) PBS/Tween four times for 5 minutes each.

- Horseradish peroxidase (HRP)-conjugated goat antibody is added and incubated for 1 hour.
- Washing is done five times with PBS/Tween and followed twice with PBS alone.
- Substrate solution of DAB or any other substrate (i.e., TMB) compatible with HRP is added and allowed to act for 5–10 minutes, after which the bands are visualized. The strips are then washed with distilled water.
- Photography can be done when the strips are still wet. Dried strips should be rewetted for any photography.

  Note: Buffer containing sodium azide should not be used for HRP, since it inhibits the enzyme.

  Results and interpretation: The interpretation of the EITB bands differs depending on the type of antigen used. The interpretation guidelines are summarized in Table 12.1. Representative EITB test runs with interpretation are shown in Figure 12.2.

Quality assurance: The following controls should be included while setting up the gel run, which should be transferred to the membrane used for testing.

Known positive control: It is used to check that the procedure and reagents are optimized and working.

Known negative control: It is used to check for nonspecific binding of antibody and false-positive results.

Loading control: Actin, GAPDH, tubulin, and others can be used to confirm that protein loading is the same across the gel and ensure the reliability of data.

TABLE 12.1
Interpretation Guidelines for EITB for Cysticercosis Including Neurocysticercosis

SN	Type of Antigen	Banding Patterns	Remarks
1.	Cyst fluid antigen (Arora et al. 2020)	15 kDa band 35 kDa band	Sensitivity and specificity >91% Multiple cysts
2.	ES Antigen of *Taenia crassiceps* (Espindola 2002)	18 and 14 kDa peptides of ES24 18, 14, and 30kDa peptides ofES48	Faint bands Prominent bands
3.	LL-GP antigen (Tsang et al. 1989)	GP50, GP42–39, GP24, GP21, GP18, GP14, and GP13	No single band alone is recognized by sera from all the patients. However, the 18 and 14kDa bands are most common ones identified in serum. All seven glycoprotein bands are uniquely specific for cysticercosis. Antibody pattern usually appears consecutively, starting with the heaviest protein GP50 band and concluding with the smallest GP13 band. Antibody pattern of disappearance during the cyst degeneration occurs in reverse order. GP50 are the first to appear and last the longest. Responses against any band from the 8 kDa family (low-molecular-weight bands GP21, GP18, GP14, and GP13) are commonly related to active infections or high antigen burden (multiple or extra parenchymal NCC) a. Responses against the T24 family (GP42–39, GP24) are heterogeneous.

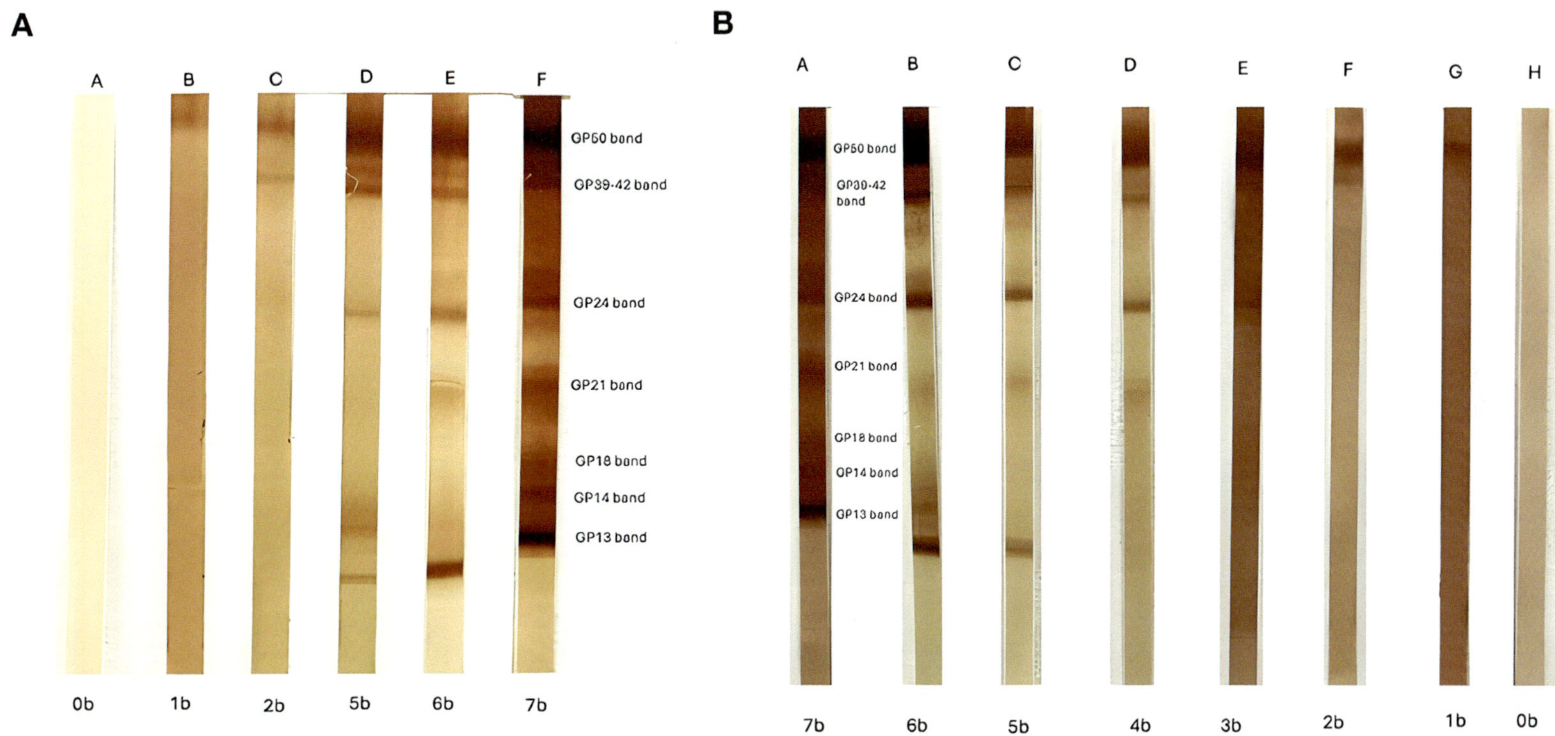

FIGURE 12.2 (A) Enzyme-linked immunoelectrotransfer blot (EITB) appearance antibody pattern obtained using consecutive serum samples from a pig experimentally infected with cysticercosis. The position of the seven antibody bands and the names are shown. "A" indicates the basal EITB result (0b = 0 bands) before the infection, and "B" to "F" are the EITB results during the time that the pig was followed (5 months after infection). (B) EITB disappearance antibody pattern obtained using consecutive serum samples from a neurocysticercosis patient under antiparasitic treatment. "A" indicates the initial EITB result (7b = 7 bands), and "B" to "H" are the EITB results during the time that the patient was under treatment and the infection was finally resolved.

TABLE 12.2
Troubleshooting in Immunoblot Technique

Error/Abnormality	Type of Error	Possible Reason	Remedy
Abnormal bands	Unexpected bands	Protease degradation	Use fresh sample that had been kept on ice or change the antibody
	High position of the protein	Intact quaternary structure	Reheat the sample
	Blurred bands	High voltage or air bubbles during transfer	Use lower voltage, and position the stack properly
Absence of bands		Use of improper primary/secondary antibody	Check the antibodies used
		Low concentration of antigen or antibody	Transfer/electrophoresis steps need checking Too much dilution of sample to be avoided
		Excess washing	Optimum washing
Weak bands		Low concentration of antigen or antibody	Transfer/electrophoresis steps need checking Too much dilution of sample to be avoided
		Less exposure time	Increase incubation time
		Non-fat dry milk masking the antigen	Use BSA instead of dry milk
High background on the blot		High antibody concentration	Use diluted sample
		Old buffers	Use freshly prepared buffers
		Less washing	Increase washing time
		Too much exposure time	Decrease incubation time
Patchy or uneven spots		Improper transfer	Avoid air bubbles between gel and membrane during transfer
		Antibodies binding to blocking agent	Change or filter the blocking agent
		Aggregation of secondary antibodies	Centrifuge and filter

Troubleshooting in immunoblot assays: Table 12.2 provides a list of important errors and their remedial measures. For details of errors and troubleshooting related to band appearance, blot background, signal strength, band size and pattern, and electrophoresis-related errors with their solutions, readers can consult the following sources:

Troubleshooting Western Blots with the Western Blot Doctor™ of Bio-Rad, available at: https://www.bio-rad.com/en-in/applications-technologies/troubleshooting-western-blots-with-western-blot-doctor?ID=MIW4HR15

Western Blot Troubleshooting by Thermo Fisher, available at: https://www.thermofisher.com/in/en/home/life-science/protein-biology/protein-biology-learning-center/protein-gel-electrophoresis-information/western-blot-troubleshooting.html

Protein gel electrophoreses tips and troubleshooting guide, available at: https://www.gomeslab.net/_files/ugd/00f296_e7f26e5995454a0f96ee3fa72e70bd0f.pdf

Commercially Available Immunoblot Kits for Parasitic Infections

Experimental immunoblots have been designed by various practitioners focusing mainly on cysticercosis. However, reports exist about their application in the diagnosis of echinococcosis, paragonimiasis, malaria parasites, toxoplasmosis, and other parasitic diseases. Prepared Western blot strips are available commercially (see Table 12.3). The list of suppliers can be found in Appendix B.

TABLE 12.3
Available Commercial Immunoblot Kits for Some Important Parasitic Infections

Protozoan Infections	Helminthic Infections
Toxoplasma gondii	*Taenia solium* (cysticercosis)
Trypanosoma cruzi	*Echinococcus granulosus*
Leishmania spp.	*Fasciola hepatica*
	Toxocara spp.
	Schistosoma spp.
	Trichinella spp.

CONCLUSION

Immunoblot assays are powerful test platforms with enhanced specificity, since the interpretation relies on recognition of multiple antigen bands and their patterns unique for a specific genus or species of parasite. There are multiple modifications of immunoblots, of which the Western blot test is the most prominent one. Several commercial tests are available for parasites in this format, but experimental immunoblot tests have also been developed for some other parasites. A careful attention to the details in the in-house preparation of the test can yield an assay of high sensitivity and specificity, which can be used for the parasite of interest.

FURTHER READINGS

Arora N, Kaur R, Rawat SS, Kumar A, Singh AK, Tripathi S, et al. Evaluation of *Taeniasolium* cyst fluidbased enzyme linked immunoelectro transfer blot for neurocysticercosis diagnosis in urban and highly endemic rural population of North India. *Clin Chim Acta.* 2020;508:16–21.

Arroyo G, Rodriguez S, Lescano AG, Alroy KA, Bustos JA, Santivanez S, et al. Antibody banding patterns of the enzyme-linked immunoelectrotransfer blot and brain imaging findings in patients with neurocysticercosis. *Clin Infect Dis.* 2018;66(2):282–288. https://doi.org/10.1093/cid/cix774

Espíndola NM, Vaz AJ, Pardini AX, Fernandes I. Excretory/secretory antigens (ES) from in-vitro cultures of *Taenia crassiceps* cysticerci, and use of an anti-ES monoclonal antibody for antigen detection in samples of cerebrospinal fluid from patients with neurocysticercosis. *Ann Trop Med Parasitol.* 2002;96:361–368.

Mahmood T, Yang P. Western blot: technique, theory, and trouble shooting. *North Am J Med Sci.* 2012;4:429–434.

Sahu PS, Parija S, Kumar D, Jayachandran S, Narayan S. Comparative profile of circulating antigenic peptides in CSF, serum, andurine from patients with neurocysticercosis diagnosed by immunoblotting. *Parasite Immunol.* 2014;36:509–521.

Sule R, Rivera G, Gomes AV. Western blotting (immunoblotting): History, theory, uses, protocol and problems. *Bio Techniques.* 2023;75: 99–114.

Toribio LM, Bustos JA, Garcia HH. From laboratory to clinical practice: An update of the immunological and molecular tools for neurocysticercosis diagnosis. *FrontParasitol.* 2024;3:1394089.

PROTOCOLS

https://www.thermofisher.com/in/en/home/life-science/protein-biology/protein-biology-learning-center/protein-biology-resource-library/pierce-protein-methods/western-blot-transfer-methods. Accessed May 28, 2025.

https://www.bio-rad.com/en-in/applications-technologies/western-blotting-transfer-techniques?ID=PQEEOP70KWE7. Accessed May 28, 2025.

https://kirschner.med.harvard.edu/files/protocols/GE_proteinelectrophoresis.pdf. Accessed May 28, 2025.
Sule S, Rivera G, Gomes AV. Detailed western blotting (immunoblotting) protocol, 2022. Available at: https://www.protocols.io/view/detailed-western-blotting-immunoblotting-protocol-yxmvmn2rng3p/v1. Accessed May 22, 2025.

13 Chemiluminescence-Based Tests

INTRODUCTION

The development of various immunoassays began with radioisotope labeling of the secondary antibody followed by the use of other non-isotope molecules such as fluorophores and enzymes. The latest development is the use of a luminescent molecule label where the antigen–antibody reaction is detected by the light emission system, termed *chemiluminescence immunoassay* (CLIA). Analogous to other immunoassays, CLIA formats may be direct or indirect, and like ELISA, it may be competitive or noncompetitive. Indirect CLIA is the commonly used method where an enzyme such as horseradish peroxidase is used.

This technology is already in wide use in virology, detection of tumor markers, and hormone assays and is making slow inroads in parasite diagnostics. Commercial CLIA kits are currently available for *Toxoplasma gondii* and *Giardia lamblia*. Meanwhile, experimental kits have been developed for the diagnosis of cryptosporidiosis, echinococcosis, and several other conditions.

PRINCIPLE OF CHEMILUMINESCENCE IMMUNOASSAY

There are certain molecules, called luminescent molecules, that emit light in the visible or near-visible wavelengths when excited. This phenomenon takes place when electrons are transferred from a higher excited state to the ground state. The potential energy contained is converted to visible light. Luminol ($C_8H_7N_3O_2$) and luciferin ($C_{11}H_8N_2O_3S_2$) are molecules that can emit light directly—a phenomenon termed *chemiluminescence*.

Chemiluminescence immunoassay combines the techniques of chemiluminescence and immunoassay. In the direct chemiluminescence method, a specific molecule is chemically activated by oxidation with the production of excited intermediates, which then release energy by light emission. In the indirect chemiluminescence method, the specific molecule, after chemical excitation, transfers the energy by chemiluminescent resonance energy transfer to an adjacent fluorophore molecule. This one is, in turn, excited and then releases some of its energy by emitting light (Figure 13.1).

In effect, the CLIA uses an enzyme and a light-producing substrate similar to the substrate in ELISA or fluorescent molecule in fluorescent antibody techniques. In the indirect method, the enzyme–substrate combinations commonly used are alkaline phosphatase with adamantyl 1,2-dioxetane aryl phosphate (AMPPD) and horseradish peroxidase with luminol or its derivatives such as isoluminol. The intensity of the light emitted is measured with using a luminometer, and the analyte is quantified. The luminometer contains a photomultiplier tube that detects the released photon, which is then amplified to high-energy current and converted to digital current by the amplifier. This is then transmitted to the computer that calculates the luminescence and quantifies it.

TYPES OF CHEMILUMINESCENCE IMMUNOASSAYS

The various types of CLIA parallel the different formats seen in ELISA. Similar to ELISA, CLIA can be direct, indirect, competitive, or sandwich type. Based on the formats, CLIA can be used for detection of antigen or antibody. Antibody detection can be quantitative or qualitative.

DOI: 10.1201/9781003644224-16

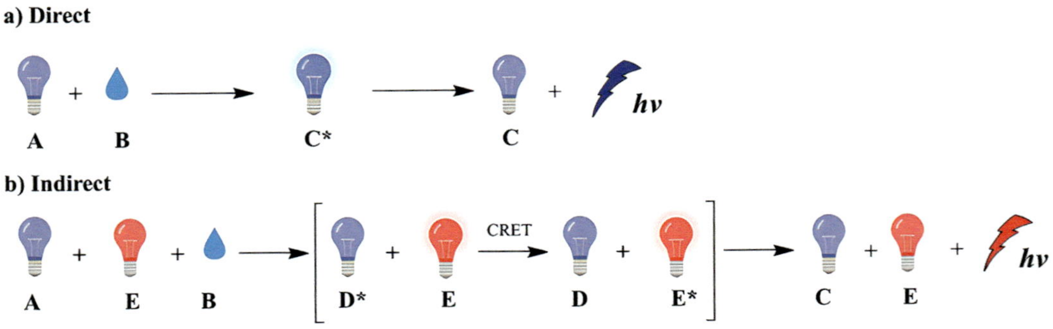

FIGURE 13.1 Principle of the direct and indirect chemiluminescence. A, Luminescent molecule; B, oxidizer; C*, excited intermediary; C, ground state, hv: photon energy release (light emission); D*, excited intermediary; D, ground state; E, adjacent fluorophore; E*, excited fluorophore; CRET, chemiluminescent resonance energy transfer.

(Reproduced from Tzani MA, Gioftsidou DK, Kallitsakis MG, Pliatsios NV, Kalogiouri NP, Angaridis PA, et al. Direct and indirect chemiluminescence: reactions, mechanisms and challenges. *Molecules*. 2021;26:7664.)

Competitive CLIA

In this type, a fixed amount of chemiluminescent labeled antigen competes with unlabeled antigen (patient's sample) for a limited number of antibody-binding sites. Here, the intensity of light emitted is inversely proportional to the amount of antigen present.

Sandwich CLIA

In this format, the antibody is fixed to a solid support, and the patient's antigen binds to the antibody. This antigen–antibody complex is then detected using a chemiluminescent labeled antibody. The light emission will be directly proportional to the amount of analyte.

Applications of Chemiluminescence Immunoassays

Chemiluminescence immunoassays are widely applied in clinical and biomedical fields due to their high sensitivity and specificity. They are commonly used in clinical diagnostics for the detection and quantification of hormones, tumor markers, cardiac markers, and antigens or antibodies associated with infectious diseases.

The CLIA play an important role in the diagnosis and monitoring of infectious diseases, with a growing significance in the detection of parasitic infections. In parasitology, CLIA is used to identify specific antigens or antibodies related to parasites such as *Plasmodium* (malaria), *Toxoplasma gondii*, *Giardia intestinalis*, and *Trypanosoma* species. The high sensitivity of CLIA allows early detection of infections, even when the parasite load is low or clinical symptoms are mild, which is particularly useful in chronic or asymptomatic cases. These assays also support large-scale screening, disease surveillance, and treatment monitoring due to their rapid turnaround time and compatibility with automated laboratory systems. Overall, CLIA has distinct advantages including rapid analysis, a wide dynamic range, and suitability for automation.

IMPORTANT CONSIDERATIONS IN DESIGNING CLIA

Chemiluminescent Labels

There are various classes of labels used in CLIA, which may be broadly classified into direct labels and indirect labels. In the indirect label, an enzyme is used.

Direct labels include the following:

a. *Luminol and related compounds*
Luminol was the first molecule used in immunoassays. It is an oxidizer like hydrogen peroxide and a catalyst like ferricyanide in an alkaline medium. These are necessary to oxidize the luminal to an excited-state intermediary, which then comes to the ground state with emission of a blue light. Isoluminol, the 6-amino isomer of luminol, and its amino substituted analogue, N-(4-aminobutyl)-N-ethylisoluminol (ABEI), are preferred over luminol because of their more efficient light emission potentials. Detection sensitivity using these molecules is 1–10 picomoles/L of the analyte.

b. *Acridinium esters and related compounds*
Acridinium phenyl esters emit an intense blue flash of light lasting less than 5 seconds when a mixture of sodium hydroxide and hydrogen peroxide is added. The pH for the reaction should be 12.0–13.0. Acridinium carboxamide analogues can also be used. The detection limit using this molecule is 5 famtomoles/L of the analyte.

Indirect labels include chemiluminescence enzymes such as the following:

a. *Dioxetanes and alkaline phosphatase*
Aryl phosphate–containing dioxetenes (AMPPD) are the preferred chemiluminescent substrates for alkaline phosphatase enzyme because of their exquisite sensitivity in detecting as low as 1 zetamole (10^{-21} mole) of the analyte. The light emission does not occur as a flash but instead is a prolonged one, lasting more than 60 minutes. AMPPD is hydrolyzed by alkaline phosphatase enzyme to decomposing intermediates that produce the light. An incubation period is needed to achieve a steady state of light emission.

Another related dioxetane is Adamantyl 1,2-dioxetane aryl galactoside substrate (AMPGD), which requires β-galactosidase enzyme; the detection limit is 30 zetamoles.

b. *Luminol and horseradish peroxidase*
Luminol or its derivatives act as co-substrates for the horseradish peroxidase enzyme. The assay reagent includes luminol, hydrogen peroxide, and an enhancer. The enhancer (4-iodophenol or 4-hydroxycinnamic acid) increases light intensity and reduces background emission caused by hydrogen peroxide acting on luminol. As a result, the ratio of signal to background emission is increased 1,000-fold. The detection limit using this system is 5,000 zetamoles.

Enzyme labeling: Horseradish peroxidase or alkaline phosphatase enzyme labeling can be done in-house using Crowther's method (see the Further Readings section), though it is generally recommended to purchase labeled enzymes from commercial suppliers for convenience.

The following sections outline the in-house development and design of a generic protocol for sandwich CLIA for antigen detection.

DEVELOPMENT AND OPTIMIZATION OF PROTOCOL FOR SANDWICH CLIA FOR ANTIGEN DETECTION

The following procedure describes generic detection of antigen in a sample. All the reagents can be obtained commercially, including the monoclonal/polyclonal capture antibody as well as enzyme-labeled secondary antibody.

Reagents and Equipment Required

Reagents: PBS; Tween-20; casein; Tris base; luminol,4-iodophenol; hydrogen peroxide.

Equipment and labware: luminometer; microtiter plates; micropipettes.

Others: capture monoclonal or polyclonal antibody; HRP-conjugated anti-human goat antibody; standard analytes of increasing concentrations.

Design and Optimization of the Test

Opaque white microtiter plates with solid white or clear-bottomed wells are used for CLIA.

- The wells are filled with 100μL of capture antibody (poly/monoclonal) at a concentration of 1 μg/mL.
- The wells are sealed tightly and kept at room temperature for 2 hours or at 4°C overnight.
- The plate contents are discarded and washed thrice with 150 μL of PBS + 0.5% Tween 20 (PBST) washing solution.
- Blocking buffer is prepared as follows: 0.5% (w/v) casein in 50 mM Tris (pH 8.0).
- The blocking buffer is applied at a volume of 200 μL per well, and the plates are sealed and kept at room temperature for 2 hours or at 4°C overnight.
- The plates are again washed thrice with PBST.
- The detection agent is prepared by mixing equal parts of luminol and hydrogen peroxide. An enhancer like 4-iodophenol (1mM concentration when using 2.5 mM luminol and 1 mM H_2O_2) can be used. Enhanced chemiluminescence is a glow rather than a flash of light and persists for hours.

Test Procedure

- 50 μL of the sample to be analyzed is added and incubated in a sealed condition for 60 minutes at room temperature.
- The wells are washed thrice with PBST.
- 50 μL of HRP-conjugated secondary antibody (goat anti-human antibody) is added, mixed well, and incubated for 30 minutes at room temperature in a sealed condition.
- Washing steps are repeated as before.
- 50 μL of the detection reagent is added immediately, and a reading is taken within 10 minutes using a luminometer set at 425-nm wavelength.
- The relative light unit (RLU) is noted. Multiple readings should be taken until the signals plateau.

Results and interpretation: The CLIA results are always read using a luminometer, and the RLU noted.

In an experimental setup or while using a commercially available kit, a standard curve needs to be prepared with RLU, with the analyte concentrations being the two axes. To plot the curve, analytes with known lower (typically a blank with RLU=0) to higher concentrations (standards) along with the known positive and negative controls should be run along with the test samples. The curve is typically a linear one showing a direct relationship between the analyte concentration and RLU.

Quantitative results: The light intensity of the sample is located on the graph that provides the required concentration of the analyte in the test sample.

Qualitative results: Signal-to-cutoff ratio (S/CO) is the common method of interpretation where results are given as positive or negative. A predetermined threshold or cutoff value is generated using statistical tools and testing a large number of positive and negative samples. The sample's RLU is divided by the RLU of the cutoff value to interpret the test.

S/CO value > 1: Positive
S/CO value <1: Negative

TABLE 13.1
Troubleshooting in Chemiluminescence Immunoassay

Problem	Probable Cause	Solution
Weak or no signal (RLU)	Improper dilution of reagents	Pipettes to be checked
	HRP conjugate not added/ deteriorated	Ensure proper addition of reagent/ replace the conjugate
	Delay in reading plates	Plates to be read within 10 minutes in case luminol is used
	Low incubation temperature	Maintain temperature at 37ºC or as per kit instructions
High signal (RLU)	High concentration of detector antibody or antigen	Recommended dilution to be used
	Wells are contaminated	Plates to be sealed during incubation
	Reagent contamination	Color and consistency to be checked
	Longer incubation time	Follow the incubation time strictly
High variation in controls, standards, and samples	Pipetting error	Pipettes to be calibrated, no air bubbles while pipetting
	Improper washing of plates	Plates to be washed properly with automatic washers if possible
	Cross-plate contamination	Use new pipette tips when changing plates; pipette tip should not touch the plate
	Nonhomogenous samples	Thorough mixing when adding samples and reagents
Sample readings are not within range	Analyte concentration too low	Use higher volume of sample
	Analyte concentration too high	Dilute the sample
Assay drift	Interruptions during assay setup	Setup of assays should be free of interruptions and be done in the shortest possible time

Quality assurance: Known positive and negative controls must be included with each test run.

Troubleshooting: The errors in CLIA are similar to other immunoassays such as ELISA or immunoblotting techniques. Some of the important ones are summarized in Table 13.1, along with possible remedial measures.

Commercially Available CLIA for Parasitic Diseases

Commercial CLIA kits have been developed mainly for noninfectious diseases and hormone assays. Among the infectious diseases, CLIA kits are available for a certain viral marker. The range of CLIA available commercially for parasitic diseases is limited to detection of *Toxoplasma* and *Trypanosoma cruzi* antibody and *Giardia* coproantigen. Similarly, there are very few scientific reports about experimental CLIA development for parasitic diseases.

CONCLUSION

Chemiluminescent immunoassays are being increasingly used because of their accuracy and speed of testing. At present, they are commercially available for the diagnosis of only three parasitic infections, but that number is expected to increase in the near future. Laboratories may also create these tests in-house for their own purposes and for any parasite they wish to target.

FURTHER READINGS

Cricka LJ. Chemiluminiscence: Cold Spring Harbor Protocol, 2018. Available at: https://doi.org/10.1101/pdb.top098236

Crowther JR. *Methods in Molecular Biology: The ELISA Guidebook.* Totowa, NJ: Humana Press; 2009b.

Jandreski MA. Chemiluminescence technology in immunoassays. *Lab Med.* 1998;29:555–560.

Tzani MA, Gioftsidou DK, Kallitsakis MG, Pliatsios NV, Kalogiouri NP, Angaridis PA, et al. Direct and indirect chemiluminescence: reactions, mechanisms and challenges. *Molecules.* 2021;26:7664.

Unit 2

Miscellaneous Tests

14 Immunohistochemical Techniques

INTRODUCTION

Histopathological examination of tissue sections obtained from the body is traditionally carried out using the classical hematoxylin and eosin (H&E) stain along with some other staining methods in special circumstances. Infectious agents can be visualized by routine staining, but in certain circumstances the tissue sections stained by these methods do not provide enough details for a final diagnosis. Problems arise when the disease-causing agent is hard to distinguish from normal or abnormal cellular structures. Immunohistochemistry (IHC), which combines the principles of histology, immunology, and biochemistry, is of immense help in such situations. In IHC, specific antigens in the tissue are detected using monoclonal or polyclonal antibody employing an enzymatic or fluorogenic label.

The genesis of IHC may be traced back to the publication by Albert Coons and his coworkers in 1942, who demonstrated pneumococcal antigens in tissues using fluorescent antibody. Since then, the IHC has improved diagnostic sensitivity and specificity compared with routine histopathological technique. The procedure needs no sophisticated instruments except an ordinary optical or a fluorescent microscope.

PRINCIPLE OF IMMUNOHISTOCHEMISTRY

The underlying concept of IHC is quite simple. The antigen present in the tissue is detected using specific antibodies. The visualization of the antigen–antibody reaction is done using an enzyme–substrate system (colorimetric) similar to ELISA or using a fluorescent label.

Tissue samples are fixed, commonly using chemicals such as formalin. Either a frozen or a paraffin-embedded section is used. The antigenic sites get concealed by formalin treatment, hence an antigen retrieval step is performed using either heat or chemicals like proteinase K. This helps unmask the antigenic sites. To allow detector antibodies to access the interior of the tissue, a permeabilization step may be performed, and a blocking agent is used to reduce background staining. The binding of the antibody with the target antigen is visualized either by immunoenzymatic method or immunofluorescence. Detection methods can be either direct or indirect, depending on the relative abundance of the antigen. If the antigen is available in sufficient amounts, the direct method is used where a labeled antibody is directly applied. The indirect method is a more sensitive technique used in situations where the amount of antigen may be low. Here a secondary labeled antibody is used.

TYPES OF IMMUNOHISTOCHEMISTRY

Depending on the type of label used, the IHC techniques can be of immunofluorescence and immunoenzymatic types (Figure 14.1).

IMMUNOFLUORESCENCE IHC

This was the original format developed by Coons and colleagues in 1942, and it became popular in the 1960s. The principle is the same as that described in the previous chapter on immunofluorescence methods. Fluorescein, rhodamine, or auramine are the commonly employed fluorophores used.

DOI: 10.1201/9781003644224-18

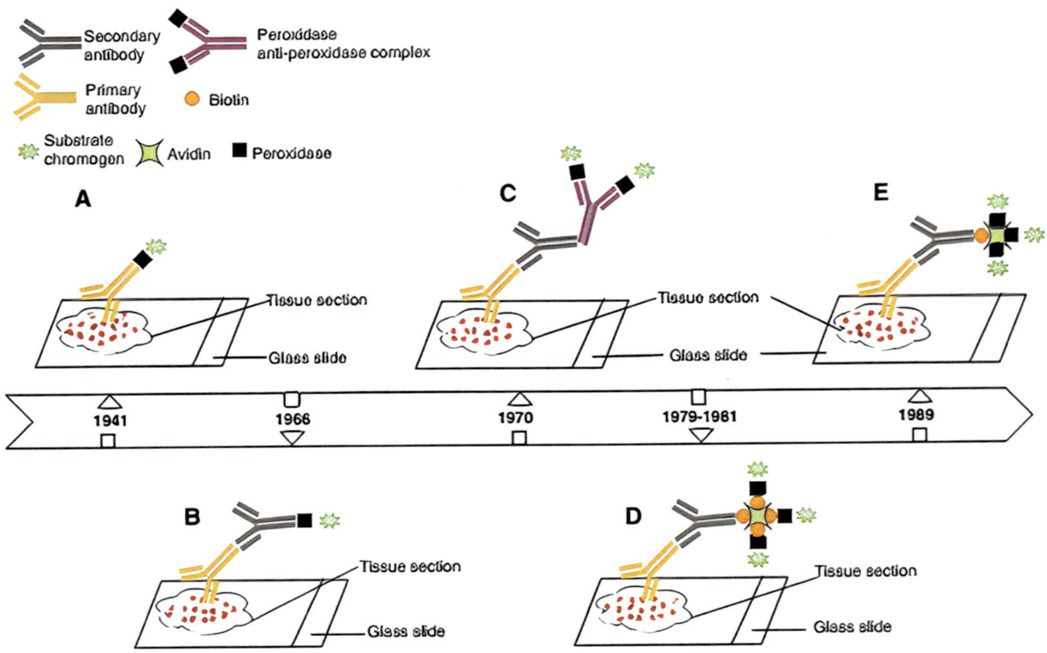

FIGURE 14.1 Various immunohistochemical techniques and their year of introduction. (A) Direct method, (B) indirect method, (C) PAP method, (D) ABC method, (E) LAB method.

(Photo from Oumarou Hama H, Aboudharam G, Barbieri R, Lepidi H, Drancourt M. Immunohistochemical diagnosis of human infectious diseases: a review. *Diagn Pathol.* 2022;17(1):17.)

In the direct fluorescent antibody technique, a specific antibody complementary to the antigen to be detected is labeled with the fluorescence marker and examined under a light-emitting diode or ultraviolet light microscope. Although less sensitive than the indirect method, it is a quicker procedure.

In the indirect method, the tissue sample is first treated with the first unlabeled antibody (e.g., rabbit anti-human antibody) that has specificity for the antigen. Next, the fluorescein-labeled second antibody against the first antibody is added (e.g., goat anti-rabbit antibody). This causes an increase in signal intensity. Because a greater number of labeled antibodies can bind to the antigen, the sensitivity of the technique is increased. However, this method takes more time compared with the direct type and needs checkerboard titration with both primary and secondary antibodies.

One serious drawback reported in the literature is the gradual loss of fluorescence over time, making it necessary to examine the preparation immediately. Although it is partly true that these preparations cannot be archived for many years like routinely stained histopathology slides, they can be stored for months to a few years without appreciable loss of fluorescence if appropriately pretreated. This can be done by using the mounting solution polyvinyl alcohol (Evanol) with DABCO (an anti-quenching/anti-fade agent) and glycerol. The mounted slides are stored in the refrigerator (4°C–8°C) or freezer in a sealed, light-proof container.

Immunoenzymatic IHC

In this format, an enzyme with a suitable substrate and chromogen is used, which produces a colored product. Various enzymes used include horseradish peroxidase (HRP), alkaline phosphatase (AP), glucose oxidase, and ß-d-galactosidase. HRP is by far the most commonly used enzyme which produces rapid entry and has reduced steric interference, while AP reduces nonspecific staining and prolonged detection. The chromogens include diaminobenzidine (DAB), which produces brown coloration, or 3-amino-9-ethylcarbazole (AEC), which produces a red color. DAB is more popular than AEC due its longevity and resistance to fading when exposed to light.

IMPORTANT CONSIDERATIONS IN IHC

Tissue Fixation

Both frozen sections and formalin-fixed tissues are suitable for IHC. Frozen sections fixed with ethanol/methanol/acetone provide optimal antigen preservation The use of 10% neutral-buffered formalin is adequate to preserve sufficient antigen reactivity.

Frozen tissues can be affected by the formation of ice crystals, which can affect subcellular detail and impact IHC staining. Another potential drawback of using frozen tissue is that the thickness of the section allows them to be stored at −80°C for no more than 1 year. When using formalin, it is important to remember that prolonged exposure gradually decreases antigenic reactivity. Optimum antigen preservation occurs with a 4–24-hour exposure. Too-long exposure produces a false-negative result in IHC. The formalin-fixed paraffin-embedded sections are preserved for many years at room temperature. A mercuric chloride–based fixative B5 can fix the tissues in 2–4 hours, thus preserving antigenic integrity.

To summarize, frozen sections preserve antigenicity but are less stable, while paraffin sections preserve the morphology better and are useful for long-term storage. In special circumstances, 70% ethanol or 80% methanol are used as fixatives. This preserves the antigen but can distort the nuclear or cytoplasmic details due to their dehydrating effect.

Antigen Retrieval

Use of an alcohol fixative (as in frozen sections) or polyclonal antibodies does not necessitate antigen retrieval. However, it is essential for formalin-fixed tissues, particularly when monoclonal antibody is used. Formalin fixation results in the formation of methylene bridges, which alters the structure of the protein, as well as formation of cross-links leading to masking of the antigenic sites.

The unmasking of the epitopes and restoration of antigenicity can be done by two methods:

a. *Proteolytic-Induced Epitope Retrieval (PIER)*
   Pepsin, trypsin, or proteinase K are used. Optimal retrieval conditions depend on the type of tissue, fixation, and antibody, necessitating optimization for each antigen. This method has lower success rates (some antigens are not unmasked by trypsin digestion) and can cause tissue damage.
b. *Heat-Induced Epitope Retrieval (HIER)*
   It reverses the cross-linkages and is the preferred method. The heat treatment is carried out using a citrate buffer (pH 6.0) or an EDTA buffer (pH 8.5). Microwave heating of paraffin tissue sections to 100°C is carried out in the presence of a citrate buffer. It is followed by 10 minutes at sub-boiling temperatures or 20 minutes at 100°C, followed by 2 hours at 60°C. The pressure cooker method and/or the fully automated tissue processor are seeing increased use nowadays.

Permeabilization

For intracellular and transmembrane proteins, antibodies need to reach the inside of the cell. Several techniques are available to achieve this. These include the use of detergents (Triton X-100, Tween-20) or freeze–thawing. The use of acetone or methanol as fixatives can help in direct permeabilization.

Protein Blocking

To reduce background staining due to nonspecific binding of antibodies, a blocking agent is necessary. Hydrogen peroxide inhibits endogenous peroxidase while levamisole inhibits alkaline phosphatase. Endogenous biotin is blocked with avidin. Nonspecific blocking agents most commonly used are bovine serum albumin, gelatin, or normal animal serum (e.g., goat serum if using a goat anti-mouse secondary antibody).

Antibody Selection

The primary antibody to be selected may be polyclonal or monoclonal and is produced specifically against the antigen that needs to be detected. Polyclonal antibodies are raised in animals and have multiple epitopes for the same antigen. Although they may have lower specificity and exhibit some cross-reactivity, they are broadly reactive and hence more sensitive. For parasitological IHC, polyclonal antibodies are mostly used. Monoclonal antibodies have also been employed in some studies. Monoclonal antibodies, raised from a single B-cell clone, recognize a single epitope, producing high specificity but lower sensitivity.

The choice of the labeled secondary antibodies used in indirect IHC depends on the primary antibody used, since they must bind to the primary antibody. Hence, the source of the secondary antibody needs to be different from the one for the primary antibody. For example, for a parasite, if the primary antibody has been raised in rabbit, the secondary antibody should be goat anti-rabbit antibody, which in turn should be labeled.

Detection Systems

a. *Direct or indirect detection systems*

   These can use labeled primary or secondary antibody, respectively. The indirect detection method is commonly used because it generates an amplified signal, and thus has a higher sensitivity. It also allows higher flexibility and economy, since just a few secondary antibodies can be used for a wide range of primary antibodies.

   The enzymes commonly used are horseradish peroxidase or alkaline phosphatase. For peroxidase, the chromogens that can be used include3,3′-diaminobenzidine (DAB, brown color), or 3-amino-9-ethylcarbazole (AEC, red color), or 3,3′,5,5′-tetramethylbenzidine (TMB, blue color). The corresponding chromogen for phosphatase may be Fast red or 5-bromo-4-chloro-3-indolyl phosphate and nitro blue tetrazolium (BCIP/NBT, dark purple to blue color).

b. *Peroxidase/Antiperoxidase (PAP) detection*

   This is based on the formation of a peroxidase–antiperoxidase complex that is attached to the primary antibody with the help of a bridge of secondary antibody. It involves three peroxidase molecules and two antiperoxidase antibodies based on the bivalent nature of IgG. Due to a greater number of enzyme molecules localized per antigenic site, the PAP method exhibits 100- to 1000-fold higher sensitivity than the two-step indirect method. It is, however, a time-consuming process. It also may not be very sensitive when used with formalin-fixed, paraffin-embedded tissue and a monoclonal antibody.

c. *Avidin–biotin complex (ABC) detection*

   Avidin is a tetrameric glycoprotein found in egg whites, while biotin is a small vitamin (vitamin H or B_7). Each subunit of avidin can bind to a biotin molecule with extremely high affinity, and up to four biotins can bind to the same avidin molecule. Because of the small size of biotin, it can be conjugated to other macromolecules without altering its activity.

The ABC is a three-step process. First the primary antibody binds to the antigen, followed by the attachment of biotinylated secondary antibody. In the final step, a complex of avidin and biotinylated enzyme is added, which binds with the secondary antibody. The avidin tetramer can bind to four biotinylated antibodies, in which large lattices are formed containing several reporter enzymes. This leads to signal amplification and elevated sensitivity; this method has been found to be 40 times more sensitive compared to PAP method.

The main drawback of the ABC method is its large lattice structure, which hinders efficient penetration. Other issues include the presence of endogenous biotin in tissues, which can cause background signals, and the tendency of avidin to bind nonspecifically to endogenous lectins and nucleic acids.

d. *Labeled avidin/streptavidin-biotin (LAB/LSAB) system*

In the LAB/LSAB system, the primary antibody is labeled with biotin while avidin/streptavidin is labeled with the antigen. Alternately, it can be used as a three-step process in which the secondary antibody is biotinylated like ABC system. The use of streptavidin, isolated from the *Streptomyces avidinii* bacteria, has certain advantages over avidin. Similar to avidin, streptavidin has four binding sites, but unlike avidin, it has neutral pH and is not glycosylated. Hence, the problem of nonspecific binding and background signaling encountered with avidin is eliminated. The LSAB method is 5–10 times more sensitive than the ABC method.

e. *Polymer-based method*

The polymer-conjugate technology utilizes a backbone of an inert polymer molecule like dextran or polypeptide, which can carry both antibodies and multiple enzymes. It is possible to anchor 10–20 antibodies and 40–100 peroxidase molecules on a dextran polymer. The technique is rapid, but the high molecular weight of the complex and its large size reduce sensitivity to some extent because it is difficult to penetrate the tissue. An alternative method is to polymerize enzymes in small linear molecules and attach them to the antibodies. In this way more compact antibody–enzyme complexes can be created which will increase the sensitivity, specificity, and signal intensity of the IHC.

At present, various detection systems are available, and some of the newer ones are highly sensitive, albeit expensive. Considering factors such as the type of tissue, the target antigen, and the available infrastructure, one can choose the most suitable method for IHC experiments.

The direct or indirect methods using a peroxidase enzyme and a suitable chromogen works well and remains the method of choice in most laboratories because of its simplicity and economy. The principle of the different detection systems is shown in Figure 14.1.

Design and Optimization of the Protocol for IHC

Here we describe a generalized protocol based on the methods described by Magaki et al. and Moreira et al.

Reagents and Equipment Require

Reagents: ethanol; methanol; xylene or toluene; paraffin wax; egg albumin; glycerol; trypsin; calcium chloride; sodium hydroxide; proteinase K; Tris base; Tris HCl; EDTA; sodium citrate; sodium chloride; hydrogen peroxide; avidin; powdered skimmed milk; PBS; bovine serum albumin; Tween-20; avidin–biotin complex; horseradish peroxidase; 3,3′-diaminobenzidine (DAB); hematoxylin; Tacha bluing solution; paramount; double-distilled or deionized water.

Equipment and labware: microtome; tissue processor; precoated or glass slides; ice blocks; oven; Coplin jars; humidified chamber; microwave oven or pressure cooker (ordinary or automated); ordinary light microscope.

Design and Optimization of the Test

Slide Preparation

Formalin-fixed tissue specimens are processed in a tissue processor, undergoing dehydration and clearing with varying concentrations of alcohol, followed by xylene or toluene, before being embedded in paraffin wax.

The embedded tissue is cooled on ice blocks or in a refrigerator at 4°C. Then the sectioning is done in a microtome to a thickness of 4–7 µm.

Once the sections are cut, they are floated on a warm water bath that helps remove wrinkles.

They are then mounted on adhesive-coated glass slides to ensure the tissue sections adhere properly.

Gelatine, poly-L-lysine, or silane-based precoated slides can be used.

A cost-effective alternative is egg albumin mixed with glycerol (Mayer albumin). This can be spread as a thin layer over the slide before mounting the tissue section.

The whole preparation is then gently heated to 60°C in an oven for at least 2 hours to coagulate the albumin and improve fixation. This step is bypassed if precoated slides are used.

Deparaffinization and Rehydration

The slides are deparaffinized and rehydrated by dipping into the following solutions, in sequence:

- xylene, three washes, 10 minutes each;
- 100% ethanol, two washes, 10 minutes each;
- 95% ethanol, two washes, 10 minutes each;
- 70% ethanol, two washes, 10 minutes each;
- 50% Ethanol, two washes, 10 minutes each;
- Deionized water, two washes, 5 minutes each.

Antigen Retrieval

a. *Proteolytic methods*: These include the retrieval by the enzymes trypsin or proteinase K.

   Trypsin: Working solution (0.05%) is prepared by adding trypsin stock solution (0.5%) 1 mL and calcium chloride stock solution (1%) 1 mL, to distilled water 8 mL (pH 7.8) with 1 N NaOH.

   The sections are covered with the trypsin solution for 10 minutes at 37°C in a wet chamber, followed by cooling at room temperature.

   Proteinase K: Working solution is prepared by adding proteinase K stock solution (20X) 1 mL to Tris EDTA buffer (pH 8.0) 19 mL.

   The sections are covered with the proteinase K solution for 10 minutes at 37°C in a wet chamber, followed by cooling at room temperature.

b. *Heat method*: these methods include the retrieval by microwave oven or pressure cooker

   Microwave oven: Slides are kept in 3% H_2O_2 for 5 minutes.

   - These are washed with deionized water for 5 minutes.
   - They are then immersed in a retrieval buffer (10 mM citrate solution, pH 6.0) in a tray and heated in a microwave oven at 100°C for 5 minutes.
   - These are cooled to room temperature and washed with deionized water twice.
   - Final washing of the slides is done with PBS for 5 minutes.

Pressure cooker: 500mL deionized water is added to the pressure cooker, and the heat shield is positioned at the center of the removable pan.

- The antigen buffer solution (10 mM citrate solution, pH 6.0) is prepared and added to the polypropylene staining dish to fill it up.
- The full basket is placed into dishes with the empty slots filled with blank slides and put in the decloaking chamber.
- The decloaking chamber settings are kept as follows: set point (SP1): 95°C, 30 minutes; SP2: 90°C, 10 seconds, with a SP limit of 10°C for a maximum of 10 minutes.
- The lid is secured. After the procedure is completed, the staining dish is kept on the bench to cool for 20–30 minutes.
- The slides are immersed in two changes of deionized water for 5 minutes each.
- The slides are then transferred to 0.1% TBS-Tween and kept for 5 minutes (up to 1 hour).

  Note: If an automated pressure cooker with temperature setting and timer is not available, a simple pressure cooker can be used. The antigen can be retrieved by free steaming in 10 mM citric acid solution (pH 6.0) for 3 minutes when using this method.

Blocking Steps

- To block endogenous peroxidase activity, the slides are immersed in 3% hydrogen peroxide in methanol in a Coplin jar for 5 minutes, followed by washing thrice in 0.1% TBS-Tween.
- To block endogenous biotin, avidin can be used if using the avidin–biotin complex.
- To block nonspecific ionic interactions, powdered skimmed milk, 60 g/L diluted in distilled water is applied and allowed to act for 30 minutes at room temperature.
- To block nonspecific interaction of secondary antibody, the slide is incubated with 1.0%–1.5% normal animal serum in 0.01 M PBS for 1 hour in a humid chamber at 37ºC.The species of the animal serum used depends on the host of the secondary antibody. Thus, if using goat anti-mouse secondary, goat serum should be used.

Test Procedure

Antibody staining These include primary antibody application and secondary antibody application.

Primary antibody application: Antibody concentration that will provide the lowest background staining and greatest target antigen staining can be determined by serial dilution of the primary antibody. If using a commercial antibody, the manufacturer's instructions are to be followed. If using an in-house antibody, it can be determined by checkerboard dilutions. In such cases the recommended starting dilution is 2–5 µg/mL of the antigen.

- The antibody is diluted in 0.01M PBS containing 1% bovine serum albumin, and 200 µL is applied over the tissue section.
- The mixture is incubated in a humidified chamber, first at 37ºC for 1 hour and then at 48ºC overnight.
- The preparation is washed thrice for 3 minutes each in 0.1% TBS-Tween using a different staining dish for each wash.
- *Secondary antibody application*: The specific secondary antibody should be applied following the manufacturer's specifications.
- The biotinylated secondary antibody (if using the ABC-HRP-DAB detection method) or the HRP-conjugated secondary antibody (if using the HRP-DAB detection method) is added to cover the sections and incubated at room temperature for 1 hour in a humid chamber.

- The sections are washed twice for 10 minutes each with 1% serum PBS-T.

Detection *ABC-HRP-DAB method*: The ABC-HRP reagent is added to the slide and incubated at room temperature for 1 hour followed by three washes in PBS for 10 minutes each. Appropriate dilution of the reagents is to be done per the instructions.

HRP-DAB method: No incubation needed.

- The working solution of DAB is prepared by adding DAB powder 60 mg/100 mL in 0.01 M PBS containing 1% H_2O_2.
- The DAB solution is applied to the section and allowed to act for just enough time for sufficient brown color to develop for imaging (may vary from a few seconds to 10 minutes maximum).
- Once the appropriate color intensity has developed, the slide is immediately immersed in deionized water and washed twice for 2 minutes each.
- If nuclear counterstaining is desired, hematoxylin staining can be done. The slide is kept in hematoxylin for 30 seconds, followed by multiple washes in distilled water. It is then incubated in Tacha Bluing solution for 90 seconds, followed by multiple washes in water. This step is optional.
- The slide is oven-dried for 10–15 minutes or dehydrated by passing through the following for 2 minutes each, sequentially: 95% ethanol, 100% ethanol, and xylene.
- It is mounted in paramount, a coverslip applied, and it is then observed under the microscope. The DAB reaction is permanent and stable and can be analyzed under a brightfield microscope at any time.

Results and interpretation: In the DAB method, the body components of the parasite, if present, are stained brown in color (Figure 14.2). Other tissues are stained light pink due to the hematoxylin stain.

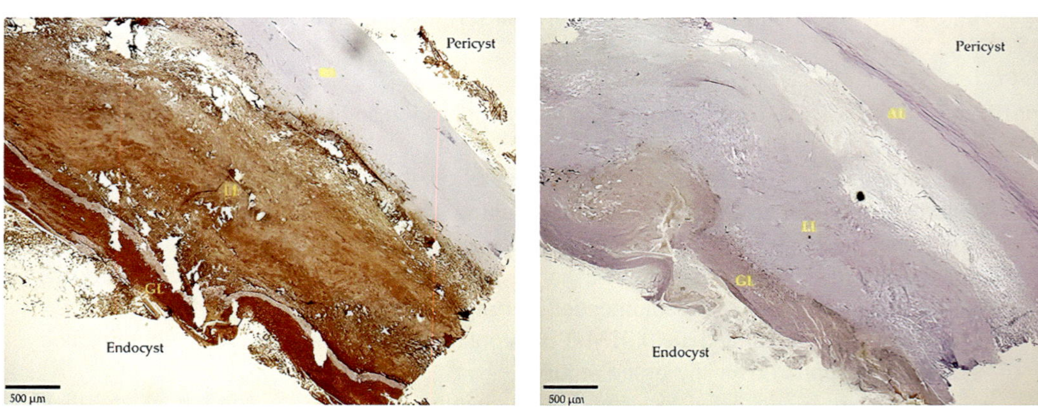

(A) (B)

FIGURE 14.2 (A) Section of hydatid cyst positive for *Echinococcus granulosus* detected by IHC using a monoclonal antibody mAbEmG3 (specific for *Echinococcus* spp.) using peroxidase-DAB staining. (B) Same section negative on staining with mAbEm2G (specific for *Echinococcus multilocularis*). AL, Adventitial layer, GL, germinal layer (magnification x4); LL, laminated layer.

(Photo from Santucciu C, Peruzzu A, Fara AM, Cossu A, Kronenberg PA, Deplazes P, Masala G. Immunohistochemistry as a reliable tool for the diagnosis of cystic echinococcosis in patients from Sardinia, Italy: a confirmatory study. *Diseases*. 2024;12(5):84.)

TABLE 14.1
Troubleshooting Guide in Immunocytochemistry

Problem	Possible Reasons	Solutions
No or weak signal	Insufficient incubation time	The incubation times for primary and/or secondary antibody must be increased
	Inappropriate primary/secondary antibody	Correct antibody must be chosen
	Inappropriate concentration of antibodies	Concentrations or dilutions must be checked
	Over- or underfixation	Appropriate concentration and time must be followed
	Insufficient deparaffinization or permeabilization	Deparaffinization steps should be checked and the correct permeabilization (0.5%–1.0% triton detergent) agent must be used
	Antigen retrieval was incorrect	Correct time/ concentration of proteolytic reagents/ temperature must be used
	The secondary antibody was not stored in the dark	Biotinylated secondary antibody should be stored in the dark
	Activity of antibodies is lost	The storage time and conditions must be checked
High background signal	Long incubation time	Recommended incubation time must be followed
	Use of polyclonal primary antibody	Monoclonal antibody can be used
	Serum blocking inadequate	Time must be increased
	Endogenous peroxidase	Adequate blocking is necessary
	Inadequate washing	More time is needed for washing
	High antibody concentration	Optimum antibody concentration must be used
	Overexposure to DAB	Slide to be rinsed with water as soon as sufficient color has developed
Nonspecific staining	The secondary antibody used was of the same species as the tissue species origin	The secondary antibody must be changed

Quality assurance: The IHC procedure is validated using appropriate positive and negative controls. The parasite with the antigen serves as the internal positive control, while a tissue with known antigen expression acts as the external positive control.

The external positive control may be stained along with the test specimen in a separate slide, or it may be integrated on the same slide with the test tissue. Similarly, a tissue section that does not express the antigen is used as the negative control.

Blocks of known positive and negative tissue sections should be preserved and should undergo the same procedure for sectioning and subsequent processing as the test tissue to achieve proper quality assurance of the IHC procedure.

Note: For IHC using a fluorophore, readers can follow the protocol developed by Im et al.

Troubleshooting in IHC: Problems encountered in IHC are mainly related to the signal strength and background staining. Table 14.1 summarizes the important problems and the possible solutions.

IHC in Parasitological Studies

Many tissue parasites can be accurately diagnosed using routine hematoxylin and eosin (H&E) or periodic acid-Schiff staining without the need for IHC. Tissue response elicited by these parasites also provides clues that help locate them and confirm the diagnosis. However, challenges may arise in distinguishing closely related parasites, and determining the exact species is not possible using these methods. For example, using histopathology, one can diagnose *Trichinella* but cannot identify the species. In addition, certain parasitic forms can resemble other organisms, including artifacts, fungal elements, or even normal tissue structures, which can lead to misdiagnosis. In addition, parasite morphology is often distorted due to tissue necrosis or autolysis. For protozoa, their small size can make a definitive diagnosis difficult.

In this context, IHC can prove to be a useful tool in tissue diagnosis of parasites. The use of IHC in parasitology remains underexploited. Currently, IHC has been investigated as a valuable tool by various researchers for the diagnosis of cutaneous and visceral leishmaniasis, toxoplasmosis, malaria, Chagas disease, and echinococcosis (Figure 14.2). In addition, IHC has also been evaluated for research purpose to study the role of various virulence factors in pathogenesis.

CONCLUSION

IHC is extensively used in histopathology to identify antigens in tissue section in several infectious and noninfectious diseases, but its application in parasitology remains limited. IHC has potential for identifying parasites in tissues that may not be readily recognized by routine staining due to distortion and sectioning. Apart from diagnosis, it serves as an important research tool to study pathogenesis of the diseases. Histopathologists have significant opportunities to use IHC as a valuable tool for identifying tissue parasites.

FURTHER READINGS

bio-techne. Immunohistochemistry (IHC). https://www.bio-techne.com/applications/imaging/immunohistochemistry. Accessed June 21, 2025.

Eyzaguirre E, Haque AK. Application of immunohistochemistry to infections. *Arch Pathol Lab Med.* 2008;132:424–431.

Gonzalez K, Calzada JE, Díaz R, Paz H, García V, Miranda A et al. Performance of immunohistochemistry as a useful tool for the diagnosis of cutaneous leishmaniasis in Panama, Central America. *Parasitol Int.* 2019;71:46–52.

Im K, Mareninov S, Diaz MFP, Yong WH. An introduction to performing immunofluorescence staining. *Methods Mol Biol.* 2019;1897:299–311.

Magaki S, Hojat SA, Wei B, So A, Yong WH. An introduction to the performance of immunohistochemistry. *Methods Mol Biol.* 2019;1897:289–298.

Mebratie DY, Dagnaw GG. Review of immunohistochemistry techniques: applications, status, and future perspectives. *Semin Diagn Pathol.* 2024;41:154–160.

Moreira MA, Luvizotto MC, Garcia JF, Corbett CE, Laurenti MD. Comparison of parasitological, immunological, and molecular methods for the diagnosis of leishmaniasis in dogs with different clinical signs. *Vet Parasitol.* 2007;145:245–252.

Oliver C, Jamur MC (eds.). *Immunocytochemical Methods and Protocols.* Totowa, NJ: Humana Press; 2010.

Oumarou Hama H, Aboudharam G, Barbieri R, Lepidi H, Drancourt M. Immunohistochemical diagnosis of human infectious diseases: A review. *Diagn Pathol.* 2022;17(1):17.

Santucciu C, Peruzzu A, Fara AM, Cossu A, Kronenberg PA, Deplazes P, Masala G. Immunohistochemistry as a reliable tool for the diagnosis of cystic echinococcosis in patients from Sardinia, Italy: Aconfirmatory study. *Diseases.* 2024;12(5):84.

15 Intradermal Skin Tests and Dye Test

INTRODUCTION

Intradermal skin tests are specialized *in vivo* tests where a small volume of antigen is injected intradermally and the local response to the injected antigen is noted. The prototype skin test is the tuberculin test used for diagnosis of tuberculosis. Nevertheless, many such tests have been described for diagnosis or epidemiological surveys for other bacteria, fungi, and parasites apart from allergen or drug (penicillin) testing in hypersensitive individuals. These tests are based on either delayed-type (Type IV) hypersensitivity reactions or Type I immediate (anaphylactic) hypersensitivity reaction. In parasitology, the best-known intradermal tests are the Casoni test for hydatid disease and the Montenegro test for cutaneous leishmaniasis. However, similar tests have also been tried for some other parasitic diseases, with variable results.

A special *in vitro* test using live *Toxoplasma* is the Sabin Feldman dye test. This test is still considered a reference test for toxoplasmosis, but it requires stringent attention to the methodology and is done in specialized laboratories. The utility of any serological test for toxoplasmosis should be assessed by comparison with the dye test to draw a meaningful conclusion.

INTRADERMAL SKIN TESTS FOR PARASITIC INFECTIONS

Several skin tests for parasitic infections have been described in the past when other sensitive serological tests were not available. Most skin tests, including Casoni test for hydatid disease, are now of historical interest due to the availability of advanced serological and nonserological techniques. Exceptions include the *Leishmanin* test, which has regained interest in recent years, and the *Onchocerca* patch test, which still retains diagnostic relevance.

CASONI TEST

This test was first described by the Italian physician Tomaso Casoni (1880–1933) in 1911 and published in 1912. Its practical implications were not clear until the 1920s, when extensive publications about its utility in diagnosing hydatid disease started appearing. Apart from humans, this test has also been used in animal diagnostics in sheep, goats, and cattle.

Principle of the Test

The Casoni test is based on Type I immediate hypersensitivity reaction. Individuals exposed to echinococcal infections become sensitized with the parasitic antigen and have IgE fixed on basophils and mast cells. Exposure to the same antigen injected intradermally results in the degranulation of these cells. This releases pharmacologically active molecules such as histamine, prostaglandins, slow-reacting substances of anaphylaxis, eosinophilic chemotactic factors, and other chemicals within a short time period. These chemical mediators subsequently produce a wheal-and-flare reaction within a few minutes on the skin, along with erythema of the surrounding area.

DOI: 10.1201/9781003644224-19

Performance of the Test

Antigen

The pooled hydatid fluid present in the uncomplicated cyst is used as the antigen. It may be of bovine or ovine origin from slaughterhouses or may be derived from intact cysts removed during surgery from humans. The cyst fluid is sterilized by Seitz filter before administration. Fluids containing scolices are particularly useful.

Test Procedure

The antigen in the amount of 0.25 mL is used for testing. The fluid is injected intradermally using a tuberculin syringe in the volar aspect of the forearm a few centimeters below the elbow joint. It produces a white raised area about 1 cm in diameter. The same volume of sterile normal saline is injected into the other forearm or a few centimeters below the antigen injection site.

Interpretation

- In a positive reaction, the test area shows an increase in swelling (wheal) with pseudopodal extensions, while the control area shows rapid subsidence of the swelling (Figure 15.1).
- It reaches a maximum dimension in 30 min, with surrounding erythematous area (flare).
- The size of the wheal is measured in two directions at right angles to each other.
- A size of the wheal at least 2.4 cm in one diameter or 2.2 cm in both diameters is considered a positive reaction.
- Doubtful reaction: Both diameters are 2.0–2.2 cm, the smaller diameter is <2.0 cm, and the larger diameter is 2.2–2.4 cm. In such cases the test is to be repeated.
- Negative reaction: The wheal is smaller than the above measurements.

Interpretation Stipulations

In patients with previous history of other helminthic infections, the test results should be disregarded. Similarly, in patients with local or generalized skin diseases, pruritus, urticaria, angioneurotic oedema, and hay fever, the interpretation requires careful consideration, since false-positive results may be present in such conditions.

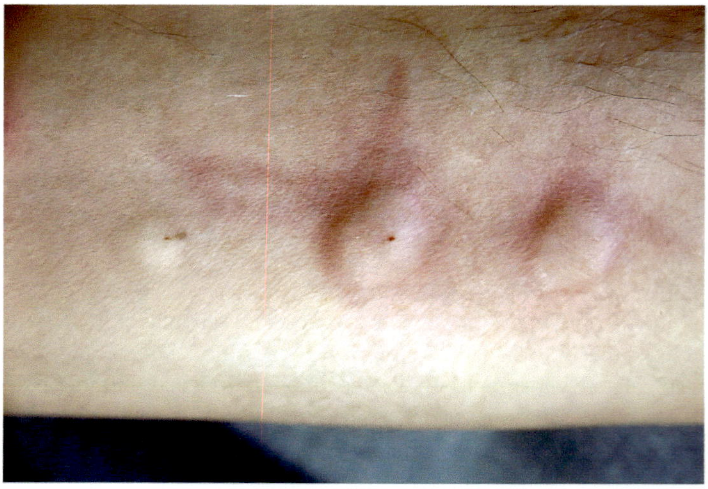

FIGURE 15.1 A positive Casoni skin test for hydatid disease.

(Photo credit: CDC / Dr. Irving G. Kagan, 1961. ID#347. Public Health Image Library, PHIL (CDC, USA).)

Precautions

There may be a remote possibility of anaphylactic shock, particularly if the test is repeated frequently. An adrenaline injection should be readily available in case of such an incidence.

Limitations of the Test

Although considered a valuable tool in the diagnosis of echinococcosis, the test has some serious drawbacks:

- *Specificity*: False positivity has been observed between 17% and 45% of the times, according to different investigators. This test shows cross-reactivity with other tissue helminthic infections, including *Schistosoma mansoni*, *Fasciola hepatica*, and *Trichinella spiralis*. In animal testing, cross-reactivity has been seen in infections with *Cysticercus pisiformis*, *Cysticercus tenuicollis*, *Taenia saginata*, and *Coenurus cerebralis*. It has been observed that high nitrogen content in the antigen (>100 μg/mL) produces false-positive results in 30–40% cases. A reduction in the nitrogen contents increases specificity. Even with the use of low-nitrogen antigen, however, a high rate of false-positive reactions has been observed with other helminthic infections.
- *Sensitivity*: The sensitivity of the test is quite high—between 80% to 95%. In general, lung cysts show lower sensitivity compared with hepatic cysts. Cerebral cysts have shown no positivity.
- *Persistence*: The test remains positive for variable time even after successful therapy. It has been found to be positive even 5 years after cyst removal.

Present Status

With the introduction of sensitive and specific serological tests and improved imaging techniques, Casoni test is rarely used as a diagnostic test for echinococcosis.

MONTENEGRO TEST (LEISHMANIN SKIN TEST)

Montenegro or leishmanin skin test is an intradermal test primarily used for diagnosis of cutaneous leishmaniasis (CL) and epidemiological surveillance for visceral leishmaniasis. Unlike Cason test, which has been relegated to history, the leishmanin skin test (LST) is still a topic of active investigation, as evident by several recent publications. The test was first reported by João Montenegro in Brazil in 1926 to be a sensitive and specific test for diagnosis of cutaneous leishmaniasis. At present, its diagnostic role has declined due to its inability to distinguish between active and cured CL, but it remains a valuable tool for epidemiological studies of CL.

Principle of the Test

This test is analogous to the tuberculin test and is based on the body's delayed-type hypersensitivity (DTH, Type IV) reaction to the antigen.

In a person exposed to leishmaniasis (past or present), injection of the test antigen or leishmanin causes activation of antigen-specific TH1 cells by the antigen-presenting cells. The TH1 cells start producing various cytokines, chief among them being proinflammatory IFN-γ. This cytokine causes local recruitment of macrophages and other inflammatory cells at the site of administration. These result in the production of an indurated raised lesion, which takes 24–48 hours to reach its maximum. An induration of at least 5 mm in diameter at 48 hours is considered as a positive reaction. It also indicates a level of intact protective cell-mediated immunity to leishmaniasis.

Protocol for *Leishmania* Antigen Preparation for Montenegro Test

Two preparations are described here, one that had been used earlier and the other described recently.

TDR Antigen

The leishmanin was prepared using killed promastigotes of *Leishmania major*.

- A specific strain of *L. major* maintained by serial passage in BALB/c mice is used.
- The organism is isolated from the skin lesions of infected mice by culture in the NNN medium.
- Parasite cultivation is carried out in RPMI culture medium prepared as follows: RPM1640, supplemented with 20% fetal calf serum (FCS), 100 U/mL penicillin, 100 μg/mL streptomycin, and 2 mM L-glutamine. Parasites from the NNN medium are transferred to RPMI and passaged three times. All incubations are carried out at 25°C.
- The stock-seed system is established using RPMI, 40% FCS, and 10% dimethyl sulfoxide containing the parasites. It is aliquoted and preserved in liquid nitrogen.
- Promastigotes are killed by exposing them to filter-sterilized 0.1% thiomersal for 30 minutes. Parasites are then harvested by centrifugation at 5,700 rpm for 30 minutes.
- Harvested parasites are washed three times using filter-sterilized and autoclaved, sterile, pyrogen-free PBS (pH 7.2–7.4) by centrifugation at 5,700 rpm for 30 minutes.
- The concentrated mixture of parasites is suspended in sterile, pyrogen-free PBS with 0.01% thiomerosal, and the parasite count is performed using a hemocytometer. The parasite concentration is adjusted to 10^7 promastigotes/mL.
- The final preparation was aliquoted in 1-mL vials and stored at 2ºC–8°C.
- *Sterility testing*: The prepared antigen is tested for bacterial and fungal contamination by culturing on trypticase soy broth, thioglycolate broth, and Sabouraud-dextrose broth for 2 weeks at 37°C.
- Pyrogen testing of antigen preparation and the PBS solution is done using the Limulus amoebocyte lysate assay.

Soluble Leishmanin Antigen

In the absence of standard leishmanin antigen, Dey et al prepared the soluble antigen using good laboratory practice (GLP), which is a prerequisite for production under good manufacturing practice (GMP) for final use in humans. The parasite strain used was *Leishmania donovani*, with the WHO code MHOM/IN/93/BI2301/LRC-751. It was isolated in India from the bone marrow of a patient with visceral Leishmaniasis.

- A frozen vial of the parasite is thawed, and the seed is prepared in M199-based growth medium. The seed is inoculated in a 2-L bioreactor containing 1 L of the M199 medium and cultivated for 96 hours.
- After 96 hours, when the promastigotes have reached a stationary phase of growth (56×10^6/mL), they are removed.
- The harvested promastigotes are washed six times using sterile PBS with 1% (w/v) glucose. Washing is done at 3,300 rpm for 5 minutes each.
- The freeze–thaw method is employed to disrupt the parasites to prepare the soluble antigen. The final washed parasite pellet is suspended in 5 volumes of sterile distilled water. Freezing is done by exposing the above preparation to liquid nitrogen, and thawing is done at 37°C in a water bath. The freeze–thaw cycle is repeated 12 times.
- The product is diluted in 10 volumes of PBS and cold centrifuged (4°C) at 10,200 rpm for 30 minutes.
- The collected supernatant is used as the soluble antigen. Its protein concentration is determined by bicinchoninic acid (BCA) protein assay. The antigen concentration is found to be 1.8 mg/mL.
- The protein concentration is adjusted to 0.5 mg/mL (50 μg/100μL) using sterile PBS supplemented with Tween-80 (0.0005% w/v) and phenol (0.28% w/v). It is sterilized using a 0.22-μm filter and stored at −80°C.

- SDS-PAGE with the extracted antigen is carried out to confirm a standard band pattern.
- Medium 199 for *Leishmania* cultivation is prepared as follows: Medium 199 (1.47 g), $NaHCO_3$ (0.22 g), heat-inactivated fetal bovine serum (5 mL), adenosine 10 mM (0.1 mM), HEPES (40 mM), L-glutamine (1 mM), gentamicin (100 µg/mL), and hemin (0.5 µg/mL). M199 and $NaHCO_3$ are dissolved in 90 mL of distilled water and pH adjusted to 7.2. Then rest of the ingredients are added, and the total volume is brought up to 100 mL. It is sterilized by passing through 0.22-µm filters.

Performance of the Test

Antigen

A suspension of killed promastigotes was originally used in several studies. However, these in-house preparations were not standardized and had variable quality, and so the results could not be reliably compared. The first standardized leishmanin was prepared by the Wellcome Trust but was discontinued in the early 1980s.

In the 1990s, the World Health Organization's Special Programme for Research and Training in Tropical Diseases (TDR) took up the task of developing a standardized antigen, and three institutes from Brazil, Italy, and Iran were asked to submit the antigen. The sources of the leishmanin were *Leishmania amazonensis* (Brazil), *Leishmania infantum* (Italy), and *Leishmania major* (Iran). Of the three, the Iranian leishmanin antigen was selected, and TDR started its distribution. However, this practice was abruptly stopped. The three laboratories continued production of the antigen in smaller scales, but all of them eventually stopped the production, the last one being the Brazilian institute in 2017.

Use of soluble antigen in place of intact promastigotes is advocated for several reasons. Disruption of the parasite for preparation of the soluble antigen makes the antigenic epitopes more easily available to be processed by antigen-presenting cells, resulting in heightened sensitivity. The soluble preparation can be easily sterilized by filtration, and the protein content can be determined, which can standardize the antigen.

Currently there is no standardized preparation of the antigen available anywhere in the world. In the protocol, we have described the methodology of two types of antigen preparations, one being the standardized Iranian antigen previously accepted by TDR and WHO and the other of a soluble antigen preparation described in a recent report.

Test Procedure

The test is performed on the volar aspect of a forearm a few centimeters below the elbow joint.

- The antigen suspension, 0.2 mL is injected intradermally to raise a small wheal.
- The suspension should contain the requisite number of promastigotes (generally 10^7/mL) when using the killed whole parasite preparation, or 25 µg of the soluble antigen.
- An equivalent volume of normal saline is injected into the other forearm.
- The injected areas should be left undisturbed and not exposed to water or soap solutions until the readings are taken.
- The ideal time for taking the reading is 48 hours after antigen administration. The size of the induration is measured by taking the mean of two perpendicular diameters, and a mean diameter of 5 mm or more is taken as positive.
- The induration boundaries are best determined using the tip of a ballpoint pen. A ballpoint pen is placed a few centimeters away from the induration (felt by the finger) and lightly moved toward the induration, drawing a line on the skin. An added pressure will be felt once the tip reaches the outer edge of the induration. The pen is lifted off, and the end of the line marks the outer boundary of the induration.

Limitation

Existing reports show no cross-reactivity with other parasitic infections. However, extensive cross-reaction exists between the various species of *Leishmania*. This may be considered advantageous, since promastigotes or soluble antigen from any species can be used as a test antigen either for CL or VL caused by any of the species without compromising the result. Thus, for example, leishmanin used by TDR, WHO was from *L. major*, but it worked equally well in other countries where the causative agents were different *Leishmania* species.

Applications of the Test

The LST has its utility in diagnostic as well as epidemiological studies, as follows (Carstens-Kass et al., 2021).

Diagnosis of CL

In the presence of clinical signs and symptoms of the disease, particularly in endemic areas, LST can be of diagnostic use as a supplemental and economical test. It also indicates an intact cell-mediated immunity.

Epidemiological Studies for CL

The LST remains positive for many years after cure has been achieved, ranging from 20 years to lifelong. It has also been observed that LST positivity may increase with age in endemic areas either due to natural cell mediated immunity or due to repeated exposure to the pathogen. This feature makes the LST a useful tool for epidemiological studies to find out the extent of the disease prevalence.

Role in Visceral Leishmaniasis (VL)

The LST remains negative in active VL due to depression of cell-mediated immunity. It becomes positive once the individual is cured and becomes positive in patients developing Post-Kala Azar Dermal Leishmaniasis (PKDL). Interestingly, LST is positive in asymptomatic VL infection similar to the finding in CL. These asymptomatic cases may serve as reservoirs of infection, and eliminating the reservoir can interrupt the chain of disease transmission. Hence, LST finds a useful role in monitoring exposure to *Leishmania* species causing VL in epidemiological studies.

Role in Vaccine Efficacy Trials

LST can be used as a good surrogate marker for testing vaccine efficacy, particularly for VL. LST can be used post-vaccination to discover the vaccine-induced immune response. There is some controversy regarding the use of LST before vaccination to identify the vaccine candidates for the trial. A negative LST in a person signifies no prior exposure to *Leishmania*, making that person the ideal candidate for a vaccine trial. However, there remains a possibility of sensitization of the individual with the test dose so that post-vaccination LST testing to determine vaccine efficacy may generate a false-positive interpretation.

ONCHOCERCA SKIN TEST

Onchocerca Diethyl Carbamazine (DEC) Patch and Mazzotti's Reaction

Mazzotti in 1948 showed itching, rash, and lymphadenitis in onchocerciasis-affected individuals after administration of 50 mg DEC, although a similar reaction was observed with a *Mansonella streprocerca* infection. The reaction results from the destruction of microfilariae by DEC. Its severity, especially in heavily infected individuals, and the associated risk of ocular damage necessitated the adoption of a safer test.

The DEC patch test is used by the Onchocerciasis Control Programme (OCP) of West Africa. It consists of a filter paper patch containing 20% w/v DEC in a commercially available skin lotion applied to a perforated adhesive piece of plaster. The patch is applied on the left iliac crest after cleaning the area with 70% methylated spirit. The patch is removed after 24 hours, and the results are scored according to the OCP patch manual. The presence of discrete, well-formed nodules ranging in number from 1 (score of 1) through 4–8 (score of 2) to >8 (score of 3), or coarse or confluent papules or extensive oedema with *peau d'orange* appearance (score of 4), is taken as positive.

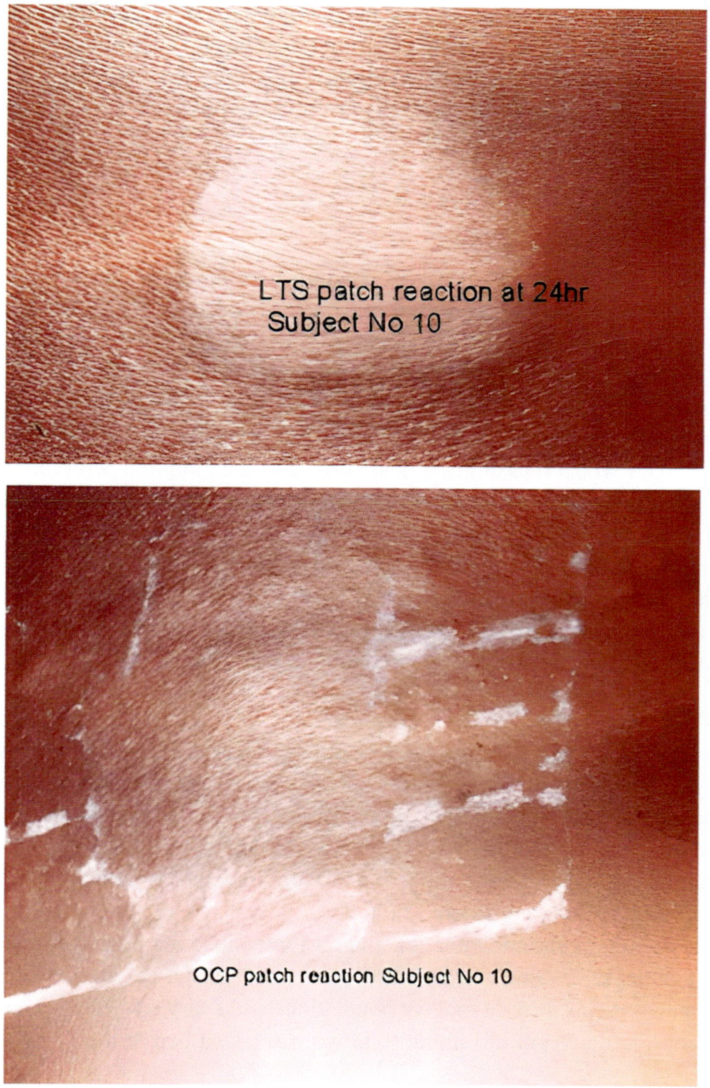

FIGURE 15.2 *Onchocerca* DEC patch test.

(Photo from Awadzi K, Opoku NO, Attah SK, Lazdins-Helds JK, Kuesel AC. Diagnosis of *O. volvulus* infection via skin exposure to diethylcarbamazine: clinical evaluation of a transdermal delivery technology-based patch. *Parasites Vectors*. 2015;8:515.) (This is an open access article distributed under the terms of the Creative Commons Attribution IGO License (http://creativecommons.org/licenses/by/3.0/igo/legalcode), which permits unrestricted use, distribution, and reproduction in any medium, provided the original work is properly cited.)

The most frequent Mazzotti reaction observed with OCP patch is mild itching. It is an inexpensive, noninvasive method that is specific for *Onchocerca volvulus*. A transdermal delivery technology patch has also been tested comparable with the OCP patch with lesser duration and intensity of itching manifestations (Figure 15.2).

Intradermal Skin Test

Cifferi and coworkers reported the use of adult *Onchocerca* antigen for skin testing in 1965. Later on, in 1970, Ulrich used a microfilarial antigen. Subsequently, a crude extract preparation was used taking the microfilaria from the skin nodules. The larval forms are homogenized and centrifuged, and the supernatant is collected and its protein content adjusted. The final product with 20 μg of protein/mL normal saline is used as a test antigen. A volume of 0.02 mL of the antigen is injected intradermally into the forearm. A wheal measuring 56 mm^2 in individuals in an endemic area with active or past infection is considered positive. Thus, this test can be used for diagnosis of onchocerciasis as well for epidemiological studies to find the prevalence of infection.

Other Skin Tests

Amoebiasis Skin Test

This is an immediate type of hypersensitivity reaction first described by Leal in 1953, where a Seitz-filtered extract of amoebae cultured with mixed flora was used for a skin test.

Subsequently, a more specialized antigen in the form of trophozoites grown in Diamond's axenic defined medium was used as antigen. The antigen is prepared from the centrifuged supernatant of disrupted trophozoite material. The skin test is carried out with a tuberculin syringe, and 0.1 mL of the antigen preparation is injected intradermally into the flexor surface of the forearm, along with saline control at another site. A wheal measuring at least 12 cm in the long axis with multiple pseudopodia and defined zone of erythema extending 15–60 mm across is considered positive. The changes appear within 2–5 min and reach their maximum at 15 minutes after the antigen administration. The wheal began subsiding in 15–20 minutes and resolved in 1–3 hours.

This test did not achieve widespread acceptance, since it is positive in active cases (liver abscess, amoebic colitis), asymptomatic cyst passers, past infection, and in individuals residing in an endemic area. Nevertheless, it may still have a role in epidemiological studies.

Toxoplasmin Test

J. K. Frenkel first described the skin test for detecting the past and latent toxoplasma infection in 1948. It is a delayed-type hypersensitivity reaction. In a subsequent version of the test, toxoplasma antigen was prepared from culture supernatants, purified, and standardized. The volume of 0.1 mL of the antigen is injected intradermally, and the results noted at 24 and 48 hours. An area of erythema and a central wheal measuring more than 10 mm in diameter constitute a positive reaction, while a diameter of 5–10 mm is considered doubtful.

The test is positive both in chronic infection and past infections but negative in acute toxoplasmosis. Various investigators have conclusively ruled out its use for the diagnosis of toxoplasmosis but have noted that it may serve as a crude tool for population-based surveys. There is some evidence that this test may be useful in detection of immunity in women tested before their first pregnancy.

Cruzin Test

Cruzin test is a skin test used for diagnosis of American trypanosomiasis. This is another test of historical importance where a purified antigen from *Trypanosoma cruzi* (T12E) was employed to detect delayed-type hypersensitivity reaction in infected subjects. To elicit a response, 50 µg of the antigen was used. It was found to be positive in Chagas disease, and its specificity was similar to the immunofluorescence test, but its sensitivity was inferior to the indirect hemagglutination test.

Ascaris Skin Test

The *Ascaris* skin test is an intradermal test based on a delayed-type hypersensitivity (Type IV) reaction to *Ascaris* antigens. It was historically used to detect prior exposure or sensitization to *Ascaris lumbricoides*. The antigen used is a crude extract prepared from adult *Ascaris* worms. Because of its crude nature, the antigen lacks standardization and may contain cross-reacting components.

Ascaris *Antigen and Test Procedure*

Live adult worms are collected from infected individuals and immediately placed in normal saline and kept at 37°C. The worms are triturated with Evans solution and petroleum ether. The preparation is centrifuged, and the ether decanted off. It is kept at 5°C for 72 hours and then filtered.

The extract is then dialyzed against pH 7.0 buffer and is diluted 1:100 in normal saline with thiomersal (1:10,000 dilutions) and finally sterilized with a Seitz filter. A volume of 0.1 mL of this antigen at a dilution of 1:1,000 is injected intradermally at the anterior surface of forearm to produce a bleb of about 6 mm in diameter. The readings are taken at 20 minutes and at 24 hours post injection.

Reactions in which the bleb expands to more than 10 mm in diameter are considered positive. The immediate reaction is found to be positive in patients who were presently infected or those who had had an infection in the past. The late reaction is positive in those with an existing infection but not in those with a past infection. Interestingly, the test was strongly positive in both early and late phases in diabetic patients.

The test has been used as an epidemiological tool to assess exposure to the *Ascaris* infection in communities. It is not recommended for individual diagnosis of ascariasis. It is simple and inexpensive to perform. Poor specificity of the test due to cross-reactivity with other helminth infections and lack of standardization in antigen preparation and interpretation are its major limitations. Hence, because of its limited diagnostic value, the *Ascaris* skin test is now mainly of historical interest.

Trichinella Skin Test (Bachman Test)

G. W. Bachman in 1928 first described a skin test for diagnosis of trichinosis using a saline solution of an antigen of *Trichinella* larva. It was extracted using Coca solution (5 g NaCl, 2.75 g Na_2HCO_3, and 4g phenol in 1L of distilled water). Later on, several workers used the test with different modifications for antigen extraction.

The test is based on an immediate hypersensitivity reaction after 0.1 mL of 1:10,000 dilution of the antigen solution is injected intradermally in the flexor aspect of forearm while the same volume of control saline is injected in the other arm. A wheal of 7 mm or more in diameter or an erythematous zone up to 20 mm developing within 15–20 minutes is considered positive.

Previous investigators also observed the development of a delayed reaction in some individuals, which became prominent after 24 hours. These subjects developed a slightly swollen, tender, reddened area 1–3 cm in diameter similar to a tuberculin test, although they did not exhibit any immediate reactivity. It was concluded that the delayed reaction was seen in early cases as well as in chronic long-standing quiescent cases, while the immediate type becomes positive from the second to third week of infection.

Paragonimus Skin Test

The intradermal test for paragonimiasis had been used in the past in China, Japan, and India. It is based on an immediate hypersensitivity reaction.

Various antigens have been used by different workers, such as saline extracts or purified antigens, and test doses have varied from 0.01 to 0.1 mL. The wheal diameter is measured immediately and again after 15 minutes, and a difference of ≥ 5 mm with erythema and pseudopodia is considered a positive test. In certain positive individuals, a large zone of erythema without appreciable wheal may occur. A negative skin test rules out paragonimiasis while a positive test may indicate either an ongoing infection or a past infection that has completely resolved. Cross-reactivity with clonorchiasis and schistosomiasis had been observed when using the crude antigen but not with the purified one.

The persistence of the test reactivity even after 10–20 years makes it an important tool for epidemiological studies to find the prevalence of infection. It is also considered a useful test to distinguish between paragonimiasis and tuberculosis in areas where these diseases are co-endemic.

Schistosoma Skin Test (Fairley Test)

Hamilton Fairley in Australia in 1927 first reported a skin test for diagnosis of schistosomiasis. He used an antigen made from dried, powdered livers of snails (*Planorbis exustus*) infected with cercariae of *Schistosoma spindalis* extracted from a goat and dissolved in saline. This antigen produced positive reactions in all cases of bilharziasis.

Later, various antigen preparations were developed from adult worms, cercaria, and eggs using various species of *Schistosoma* (*Schistosoma bovis*, *Schistosoma mansoni*, *Schistosoma japonicum*) and even non-*Schistosoma* species (*Fasciola hepatica*, *Fasciola gigantica*, *Planaria maculata*, etc.) and tried by numerous practitioners. One antigen, considered the "WHO Reference Skin Test Antigen," was a sterile, acid-soluble protein fraction from *S. mansoni* adult worm containing 0.028 mg N/mL in 1:5,000 Merthiolate. Pellegrino in 1958 attempted to standardize the test in terms of different antigens and criteria for its interpretation.

Schistosoma skin test is based on an immediate hypersensitivity reaction. When 0.05 mL of the antigen is injected subcutaneously, a wheal with an area of 1.2 cm^2 or more after 15 minutes is considered a positive test. Wheals measuring 1.0–1.1 cm^2 are regarded as doubtful, while those smaller than 0.9 cm^2 are considered negative. The test is positive in patients with schistosomiasis and remains so for a few years after apparent cure. It is found to be positive in residents of endemic areas. The proportion increased with the age of the subjects. The variable sensitivity and specificity of the test are partly due to nonstandardized antigen sources. The discovery of better serological and other tests has made this test nearly obsolete.

Sabin–Feldman Dye Test for Toxoplasmosis

In 1948, Sabin and Feldman first described a test to detect specific antibodies to *Toxoplasma gondii* and to differentiate between acute and latent infections. Soon the procedure and interpretation of the test were standardized, followed by modification of the test to a microtiter plate method. The development of newer serological tests such as immunofluorescence, ELISA, and immunoblot methods diminished the use of the dye test. The test is still considered the gold standard for all serological tests. However, due to the expertise required and the use of live organisms, this test is now performed only in specialized reference laboratories.

Principle of the Test

The test is based on the presence of neutralizing or cytoplasm-modifying antibodies in the patient. It can detect the total antibody present (IgG, IgM, IgA). When the serum of the patient along with accessory factor (complement, present in normal human serum) is added to a suspension of live tachyzoites of the parasite, it causes membrane damage, which can be demonstrated by adding the methylene blue dye. The damaged parasites do not take up the stain and remain colorless when seen under the microscope, indicating a positive test (Figure 15.3). The antibody titer is calculated by

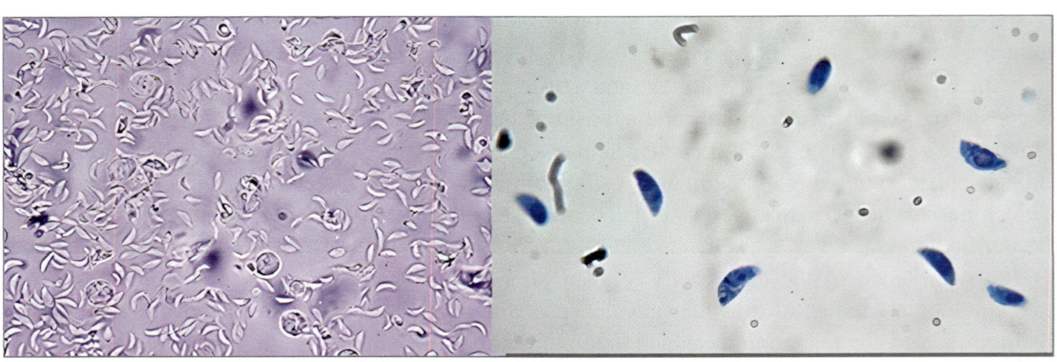

FIGURE 15.3 Sabin–Feldman dye test. Left: A positive test. Right: A negative test.

(Picture credit: YouTube. Krzyściak P, Salamon D, Malska A, Papir B. Footage of live *Toxoplasma gondii* parasite. *World J Med Imag Vid Cases*. 2018;4:e16–17.) (The journal is Full Open Access.)

finding the ratio of live to dead parasites. The highest dilution of serum showing more than 50% unstained tachyzoites is considered the titer. In a negative test, 80%–90% of the tachyzoites are stained. A titer of 1:128 is suggested as diagnostic of acute toxoplasmosis.

Sabin–Feldman Dye Test

Antigen

Live tachyzoites of *T. gondii* are used as the antigen. There can be two sources of the antigen. In the original assay, parasites maintained in mouse peritoneal cavity were used, and later on, tissue culture maintained tachyzoites were used. Mouse-derived tachyzoites (*in vivo* culture) are more virulent and considered the ideal antigen. Cell culture–derived antigen (*in vitro* culture) is less virulent and hence safer to work with, but it sometimes generates false-negative results. When peritoneal exudate of mouse is used as antigen source, an anticoagulant such as heparin (1: 5,000) or sodium citrate solution (1%) is used to prevent coagulation.

Three *Toxoplasma* strains are commonly used: RH, BK, and, rarely, C-56. The RH strain is derived from the original Sabin isolate, while the BK strain is an isolate from the CSF of a 6-week-old infant from the Netherlands. These strains have been serially passaged in mice for many years. The BK strain is said to produce fewer nonspecific results. Tachyzoites may be obtained from the national toxoplasma reference centers in France, the United Kingdom, and Germany or from the Palo Alto Medical Foundation Toxoplasma Serology Laboratory in the United States.

Reference Serum

The dye test must be performed using a control serum with a known antibody titer. A reliable reference serum is essential to check the test performance. Countries like United Kingdom, France, and Germany have their own national reference sera, apart from the international standard serum introduced in 2016. The Expert Committee on Biological Standardization (ECBS) of the World Health Organization (WHO) has established the serum sample previously designated as 13/132 to be the 4th International Standard for *Toxoplasma* serological testing, including the dye test. It is supplied as freeze-dried preparation 0.5 mL of pooled human plasma from six donors from France who had a recent *Toxoplasma* infection. This standard has an assigned unitage of 160 IU per ampoule for Ig by dye test and contains a high level of IgG and IgM. The international standard serum can be obtained from the National Institute for Biological Standards and Control (NIBSC) and WHO (http://www.nibsc.org/products/ordering.aspx).

Test Procedure

The test can be performed in tubes or flat-bottom microtiter plates.

- Doubling dilutions of inactivated (56°C water bath for 30 minutes) patient's serum (1:2 to 1:256) are prepared and 100 μL is dispensed in each well of the microtiter plate.
- The reference serum (positive control) and a negative saline control are run in parallel.
- To this, equal amount of normal human serum negative for toxoplasma antibody (accessory factor, source of complement) and a suspension of live tachyzoites (containing about 2×10^6 tachyzoites/mL) are added. After mixing, the plate is incubated at 37°C in a water bath for 60 minutes.
- 100 μL of saturated alkaline methylene blue solution (pH 11.0) is then added, and the whole preparation is shaken lightly.
- One drop of the mixture is taken on a microscopic slide and, after cover slip preparation, is examined under a microscope.

Results and Interpretation

- In a positive test, the tachyzoites do not take up the dye and appear colorless, while in a negative test, the tachyzoites appear blue.

- In a positive test, the titer is expressed as the highest dilution of serum at which >50% of the tachyzoites remain unstained.
- A dye test result of ≥1:16 (2 IU/mL) is considered positive.
- This test is both specific and sensitive in humans but may be unreliable in cattle and avian species, although it has been used in epidemiological studies in cats and sheep.
- It becomes positive in 1–2 weeks after infection and reaches its peak in 6–8 weeks. It declines at variable rates and usually persist for life at low levels.
- A negative DT effectively rules out past or present toxoplasmosis.
- In a small number of patients, IgG antibodies might not be detected within 2 to 3 weeks, and the test is reported to give a false-positive reaction in *Sarcocystis*, *Trichomonas vaginalis*, and *Trypanosoma lewisii* infections.

CONCLUSION

Intradermal skin tests for a variety of parasitic diseases have been described in the past, but the great majority of them have passed into oblivion because of safety concerns and poor specificity. The DEC patch test for onchocerciasis continues to be part of the control program in a few African countries and is generally safe. In recent years there has been a renewed interest in revival of the Montenegro test for leishmaniasis using improved standardized antigens. For toxoplasmosis, the Sabin–Feldman dye test remains the reference serological test and is available in a few toxoplasma reference laboratories in the United States and in some countries in Europe.

FURTHER READINGS

Alimohammadian MH, Hakimi H, Nikseresht MA. The preparation and evaluation of reference leishmanin from *Leishmania major* for use in man for diagnostic and experimental purposes. *Med J Islam Rep Iran.* 1993;7:23–28.

Anonymous. Casoni test. *Postgrad Med J.* 1946;22:203–204.

Awadzi K, Opoku NO, Attah SK, Lazdins-Helds JK, Kuesel AC. Diagnosis of *O. volvulus* infection via skin exposure to diethylcarbamazine: clinical evaluation of a transdermal delivery technology-based patch. *Parasites Vectors.* 2015;8:515.

Biagi F, Payán AM, Cuenca G, Hernández-Chiñas C. Value of a skin test and precipitin reaction in diagnosis of ascariasis. *J Infect Dis.* 1960;107:149–154.

Carstens-Kass J, Paulini K, Lypaczewski P, Matlashewski G. A review of the leishmanin skin test: A neglected test for a neglected disease. *PLoS Negl Trop Dis.* 2021;15(7):e0009531.

Dey R, Alshaweesh J, Singh KP, Lypaczewski P, Karmakar S, Klenow L et al.. Production of leishmanin skin test antigen from *Leishmania donovani* for future reintroduction in the field. *Nat Commun.* 2023;14:7028.

Hashiguchi Y, Kawabata M, Zea G, Recinos MM, Flores O. The use of an *Onchocerca volvulus* microfilarial antigen skin test in an epidemiological survey of onchocerciasis in Guatemala. *Trans R Soc Trop Med Hyg.* 1979;73:543–548.

Hiatt RA, Cline BL, Knight WB. Limitations of the intradermal test for *Schistosomiasis mansoni*: experience from epidemiologic studies in a Puerto Rican community. *Am J Trop Med Hyg.* 1978;27:535–541.

Kagan IG. A review of serological tests for the diagnosisof hydatid disease. *Bull WHO.* 1968;39:25–37.

McNaught JB, Beard RR, Myers JD. The diagnosis of trichinosis by skin and precipitin tests. *Am J Clin Pathol.* 1941;11:195–209.

Miller MJ, Scott F. The intradermal reaction in amebiasis. *Can Med Assoc J.* 1970;103:253–257.

Pellegrino J. The intradermal test in the diagnosis of bilharziasis. *Bull WHO.* 1958;18:945–961.

Reiter-Owona I, Petersen E, Joynson D, Aspöck H, Darde ML, Disko R et al. The past and present role of the Sabin-Feldman dye test in the serodiagnosis of toxoplasmosis. *Bull WHO.* 1999;77:929–935.

Rijpkema S, Hockley J, Rigsby P, Guy EC, Toxoplasma Study Group. Establishment of replacement International Standard 13/132 for human antibodies to *Toxoplasma gondii*. *Biologicals.* 2016;44:448–455.

Rougier D, Pierre Ambroise-Thomas P. Detection of toxoplasmic antibody by multipuncture skin test with excretory-secretory antigen. *Lancet.* 1985;326:121–123.

Schneider J, Goddard H, Goddard D, Heinz HJ. A toxoplasmin skin-testing survey amongst a group of South African Bantu. *S Afr Med J.* 1955;29:1162–1165.

Singh TS, Sugiyama H, Rangsiruji A. Paragonimus & paragonimiasis in India. *Indian J Med Res.* 2012;136:192–204.

Teixeira AR, Teixeira Mda G. Delayed hypersensitivity to *Trypanosoma cruzi* antigen. III–Sensitivity of the skin test with T12E antigen in the diagnosis of Chagas disease in hospitalized patients. *Rev Soc Bras Med Trop.* 1995;28:267–271.

Udonsom R, Buddhirongawatr R, Sukthana Y. Is Sabin-Feldman dye test using *T. gondii* tachyzoites from animal inoculation still the best method for detecting *Toxoplasma gondii* antibodies? *Southeast Asian J Trop Med Public Health.* 2010;41(5):1059–1064.

16 Immunodiagnostics in Animals of Zoonotic Importance

INTRODUCTION

Veterinary parasitology has traditionally relied on morphological characteristics of the organisms for diagnostic purpose in parallel with human infections. This state of affairs started to change slowly from the second half of the last century with the introduction of sensitive and specific serological tests for antigen and antibody detection (Figure 16.1). Various antigen or antibody detection kits produced commercially for diagnosis of zoonotic infections in humans have also been used in the related animals. Apart from that, experimental and commercial immunodiagnostic tests have been developed for parasitic infection of veterinary importance exclusively for use in animals. This chapter focuses primarily on the application of immunodiagnostics in animal diseases, with a few exceptions involving parasites that can cross the boundary between animals and humans.

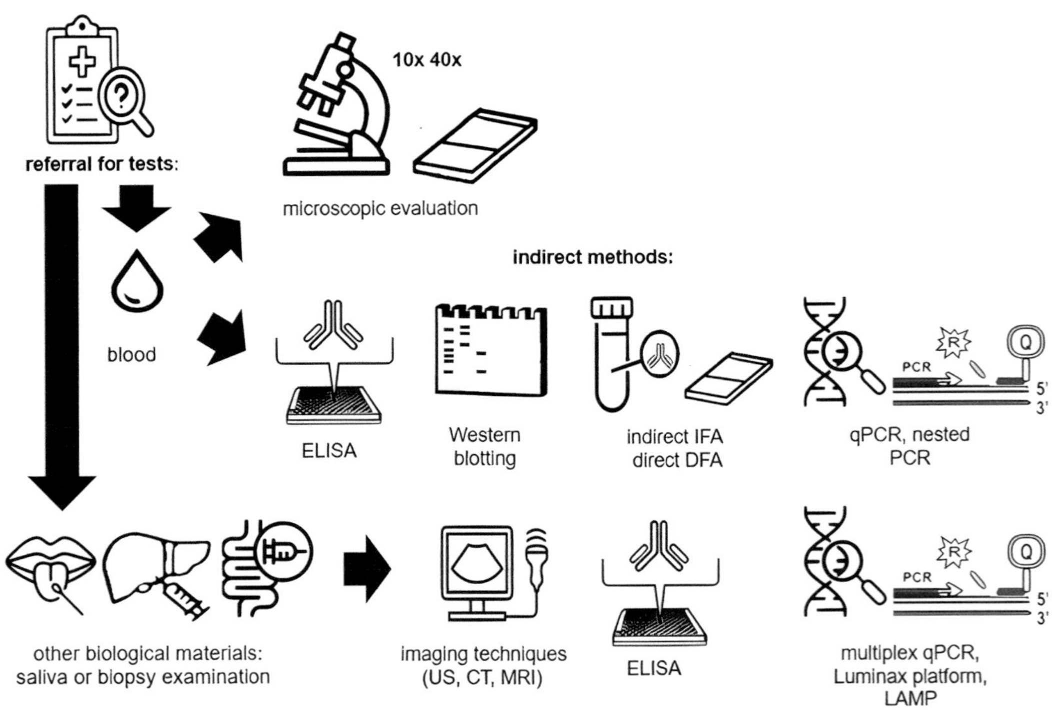

FIGURE 16.1 Diagnostic techniques (direct and indirect) for parasitic diseases in animals.

(Photo from Dąbrowska J, Groblewska M, Bendykowska M, Sikorski M, Gromadzka G. Effective laboratory diagnosis of parasitic infections of the gastrointestinal tract: where, when, how, and what should we look for? *Diagnostics*. 2024;14:2148.) (This article is an open access article distributed under the terms and conditions of the Creative Commons Attribution (CC BY) license, https://creativecommons.org/licenses/by/4.0/.)

DOI: 10.1201/9781003644224-20

IMMUNODIAGNOSTICS FOR PARASITIC INFECTIONS IN ANIMALS

Immunodiagnostics for parasitic infections in animals depends on the detection of host antibodies or parasite antigens using specific antigen–antibody reactions. These are particularly useful for epidemiological surveys, screening of carrier animals, and diagnosis of chronic or latent protozoan infections in which direct demonstration of the parasite is difficult.

IMMUNODIAGNOSTICS FOR PROTOZOAN INFECTIONS IN ANIMALS

Babesiosis

Babesiosis is caused by intracellular protozoan parasites belonging to genus *Babesia* and transmitted by a tick vector. These are important causes of disease among the livestock as well as companion animals and are seen worldwide. A few species have zoonotic potential. The geographic distribution of babesiosis has been observed to correlate with the regional prevalence of specific tick species. *Babesia bovis*, *Babesia bigemina*, and *Babesia divergens* are the species that commonly affect cattle populations and are seen in tropical and subtropical regions in Africa, Asia, South America, and Australia. In domestic dogs, *Babesia canis* (subspp. *canis, vogeli*, and *rossi*) and *Babesia gibsoni* are commonly seen, with *B. vogeli* being the most important species. It is seen worldwide, particularly in areas with high tick population.

Indirect fluorescent antibody test (IFAT) and enzyme-linked immunosorbent assay (ELISA) are serological tests commonly used in the diagnosis of babesiosis in animals. Due to the presence of a variety of *Babesia* species, cross-reactions are common in all serological tests. Pinpointing the exact species responsible for the infection may be challenging. This problem has been addressed in some ELISA formulations using recombinant antigens. These serological tests, however, are not very useful in detecting acute infections.

IFAT: This test has been used in the past for bovine babesiosis using slides prepared from blood taken directly from infected animals or culture-grown parasites. Cross-reactivity between *B. bovis* and *B. bigemina* is a common problem when using IFAT. A similar issue has also been observed in canine babesiosis.

ELISA: Parasite lysates from infected erythrocytes or recombinant antigens have been employed in designing indirect ELISA. It has largely replaced IFAT for diagnostic purposes in both bovine and canine babesiosis.

Immunochromatographic test: *B. canis–B. gibsoni* antibody detection kit is commercially available. This test uses the lateral flow immunochromatographic method in a double-layer, sandwich format using a specimen of dog serum, blood, or plasma.

Toxoplasmosis

Toxoplasmosis caused by the apicomplexan protozoan parasite *Toxoplasma gondi* is a zoonotic disease of importance. It is prevalent worldwide, particularly in countries such as Egypt, Turkey, and Ethiopia. At one time it was highly prevalent in various European countries, but the number of cases has come down in recent years. The disease is most commonly seen in cats and occasionally in dogs. The oocysts shed by an affected animal in feces are responsible for transmission of the disease to humans. It is to be noted that the cyst-passing animals are generally asymptomatic, while those with an acute infection do not shed the oocysts, since parasite multiplication occurs extra-intestinally in such cases.

Serological tests for detecting IgG and IgM are widely available in various formats including agglutination test, ELISA, IFAT, Western blot, and rapid immunochromatographic tests (ICT). IgM positivity is related to recent or active infection, while IgG signifies past infection. In cats, a very high IgM or IgG or a rising titer along with clinical illness indicates acute toxoplasmosis. Serological tests are useful for assessing an animal's susceptibility to infection and the associated risk of transmission to humans. A high IgG titer in a healthy cat indicates a previous infection, which means the animal is immune to the infection and is not excreting the oocysts. Such animals do not pose any danger to humans. A recently developed ICT using recombinant truncated surface antigen-2 has been found to be simple, rapid, and useful for field conditions.

Leishmaniasis

Dogs and other canines are highly susceptible to select *Leishmania* species, of which the most important is *Leishmania infantum*. Other species such as *Leishmania mexicana*, *Leishmania donovani*, and *Leishmania braziliensis* have been reported in dogs and sometimes in the cats in endemic regions.

The canine disease is prevalent in the Mediterranean and Balkan regions in Europe, central and southwest Asia, China, north and sub-Saharan Africa, and parts of Central and South America. These regions overlap with endemic areas of human visceral and cutaneous leishmaniasis. Although it produces mostly asymptomatic infections in the dog population, disease manifestations have been seen in immunocompromised dogs younger than 3 years and older than 8 years. Skin lesions are the most common feature, but deep organ involvement has also been noted. The majority of those affected show poor body condition: thin, cachectic, and anorexic. Renal involvement leading to failure is the most common cause of death in these animals.

Several serological tests using different types of antigens and cutoff titers have been described for diagnostic purpose as well as for epidemiological studies. These include IFAT, ELISA, and rapid ICT. Both IFAT and ELISA provide comparable results in the diagnosis of canine leishmaniasis. Vaccination for canine leishmaniasis is now practiced in many countries. A serological test needs to be performed before vaccination to know the immune status of the animal. A major limitation of vaccination is the induction of antibodies that may be detected by conventional serological assays. This may not allow differentiating between infected and vaccinated animals. As a result, vaccinated dogs may be falsely identified as having a natural infection.

IFAT: It is considered the gold standard for serological diagnosis of canine leishmaniasis. The antigen used is the intact parasite. This test has been extensively used for diagnosis and epidemiological surveys. Cross-reactivity with other *Leishmania* species and with *Trypanosoma* species remains a problem. The sensitivity of the test in symptomatic and asymptomatic dogs has been found to be 90% and 31%, respectively with specificity of 97%–98%. According to World Organization of Animal Health (WOAH, formerly OIE), in areas not endemic for *Trypanosoma cruzi*, the sensitivity and specificity of IFAT have been found to be, respectively, 96% and 98%.

ELISA: ELISA tests using soluble antigen of the parasite or sonicated parasites have been extensively used, but specificity has been found to be variable. In recent years, recombinant protein antigens are increasingly being used to have ELISA of high sensitivity and specificity. The combination of rKLO8 and rK26 proteins has shown increased sensitivity and specificity in the serodiagnosis of canine leishmaniasis. The chimeric protein "Q," containing determinants of the antigens LiP2a, LiP2b, LiPO, and LiH2A, has also been used in ELISA.

ICT: The recombinant fusion proteins rK39 and rK28 (obtained from the fusion of rK9, rK26, and rK39 from *L. donovani*) based ICT is used extensively for diagnosis in humans. It is a rapid, inexpensive test suitable for use in field conditions and available commercially from several manufacturers. This test suffers from low sensitivity (30%–70%) compared with IFAT and ELISA, and the specificity is medium to high.

Sarcocystosis (Sarcosporidiosis)

Sarcocystis spp. are intracellular coccidian parasites widely found in mammals, reptiles, and birds. It has a high prevalence in livestock animals such as cattle, goats, sheep, and pigs. Some species are also implicated in disease in pet animals such as dogs and cats. *Sarcocystis cruzi* is the most prominent cattle pathogen, apart from *Sarcocystis hirsute* and *Sarcocystis hominis*. Sheep and goat infections are commonly caused by *Sarcocystis tenella*, *Sarcocystis capracanis*, and *Sarcocystis hircicanis*. *Sarcocystis caninum* and *Sarcocystis svanai* infections are associated with dogs, while swine populations are affected by *Sarcocystis suihominis* and *Sarcocystis porcifelis*. *Sarcocystis* spp. can be found worldwide. Commonly, it causes asymptomatic infections, and tissue cysts are accidentally discovered during slaughter. Some species can cause weight loss, anemia, muscle weakness, and abortion in the affected animals. In American horses, *Sarcocystis neurona* causes a severe infection called equine protozoal myeloencephalitis.

Serological tests for diagnosis in cattle, sheep, or swine populations depend on IFAT and ELISA. Both these tests have been reported to give similar results showing no cross-reactivity with toxoplasmosis. A Dot-ELISA has been described using sonicated bradyzoites from cattle as antigen. The sensitivity and specificity of Dot-ELISA were found to be 86% and 88%, respectively, favorably comparing with IFAT (sensitivity 83%, specificity 100%).

Trypanosomiasis

Trypanosomiasis is an important zoonotic infection affecting humans and animals. A few species have geographical restriction while others spread in a more cosmopolitan fashion. Two distinct groups of infection are seen in animals, which may be transmitted to humans. These animal infections are Nagana, prevalent in Africa, and Surra, seen on various continents.

Nagana

It is commonly seen in cattle, though other animals such as horses, camels, goats, sheep, pigs, and dogs may be affected. The most important causative species are *Trypanosoma congolense* (subgenus *Nannomonas*) and *Trypanosoma vivax* (subgenus *Duttonella*), both of which do not cause infection in humans. However, two species that cause nagana in cattle and pigs and are also important human pathogens are *Trypanosoma brucei gambiense* and *Trypanosoma brucei rhodesiense*. These parasites are confined to Africa, particularly in the Tsetse fly belt lying between latitudes 15° north and 20° south and affecting 37 countries.

Immunological diagnosis is primarily based on IFAT and ELISA used for cattle. Both tests have high sensitivity and genus specificity. The antigen for ELISA is the soluble antigens of trypanosomes (*Trypanosoma congolense*, *Trypanosoma vivax*, and *Trypanosoma brucei brucei*) grown in rodents. It gives a presumptive diagnosis, since the antibodies may persist for variable periods, up to a few months after recovery from an infection. In addition, cross-reactions can occur with other trypanosomes such as *Trypanosoma theileri* (a nonpathogenic species) and *Trypanosoma evansi*. Antibody detection tests become positive 1–6 weeks after infection. Antibodies can persist for 1 to 13 months after resolution of infection (average 3 months). Depending on the geographic region (Africa or America), three ELISAs utilizing soluble antigens from whole-cell lysates of animal trypanosomes are recommended: *T. vivax*, *T. brucei*, and *T. congolense* (Savannah type). Serological tests can detect both current and past infections and provide a presumptive diagnosis for any active infection.

Surra

The causative agent of Surra is *T. evansi*, and the disease is widely prevalent in Asia, Africa, and Central and South America. The animals that get infected include cattle, buffalos, horses, and camels. Human infections have been documented in some countries, including India.

The serological tests employed for antibody detection include IFAT, ELISA, and a card agglutination test. An immune trypanolysis test has also been reported. The card test can correctly identify the infected animal and is suitable for field use. However, all serological tests suffer from the problem of low positive predictive value.

IFAT: The antigen used in IFAT is derived from infected mouse or rat blood with high parasitemia. The dried blood spot on the slide contains 5–10 *T. evansi* organisms when observed under 500x magnification.

ELISA: Two types of antigens can be used in ELISA: universal antigen and the variant surface glycoprotein (VSG) antigen. The universal soluble antigen is obtained by ultrasonication of rodent derived *T. evansi*. It is not strain specific and can detect antibodies to the parasite in various host species, cutting across geographical areas. ELISA using the *T. evansi* universal antigen show strong cross-reactivity with *T. vivax, T. congolense*, and even *T. cruzi*. The other antigen is the RoTat 1.2 (VSG based) from a *T. evansi* RoTat 1.2 clone that is a predominant antigen in *T. evansi*. It is highly specific for *T. evansi*. However, this type of ELISA may be unsuitable in regions where RoTat 1.2-negative *T. evansi* strains are present, as has been reported in Kenya. The protocols and reagents are available from the WOAH Reference Laboratory in Belgium for use in equines, camelids, and water buffaloes.

Card agglutination test for trypanosomiasis (CATT/T. evansi): This test, along with control sera, is available commercially from the WOAH laboratory in Belgium. The antigen is lyophilized stained parasites, which gives a blue-colored granular agglutination pattern with positive sera. Sensitivity varies depending on the animal species, being high in horses, moderate in camels and buffaloes, and low in cattle and pigs. Cross-reactivity has been noted with *T. evansi*, *T. equiperdum*, *T. bucei*, and *T. vivax*. Because of its high positive predictive value, it is the recommended test for surra. Another CATT/*T. evansi* version is available using the RoTat 1.2 VSG.

Immune trypanolysis test RoTat 1.2 (TL RoTat 1.2): This test detects specific antibodies directed against *T. evansi*, which can cause lysis of the live trypanosomes in the presence of complement. It can detect specific antibodies directed against variable antigen type RoTat 1.2 and is positive only with hosts that produce the trypanolytic immunoglobulins directed against RoTat 1.2. In the test, live trypanosomes are incubated with the affected animal serum for 60 minutes in the presence of a complement in the form of guineapig serum. Lysis $\geq$50% of the parasites is considered positive. It can be used as a confirmatory test.

Immunodiagnostics for Helminthic Infections in Animals

Trichinellosis

Members of the genus *Trichinella* are important zoonotic pathogens, and the disease is transmitted to humans through pigs. The genus has 13 taxa and 10 named species, of which *Trichinella spiralis* is the most common and has a wide geographic distribution. Other species such as *Trichinella britovi*, *Trichinella nativa*, *Trichinella murrelli*, and *Trichinella nelsoni* are more common in wild carnivores but sometimes seen in domestic pigs and wild boars. Affected animals normally remain asymptomatic.

Over the years, several serological tests have been developed and used for detecting trichinellosis in wild and domestic animals, particularly pigs. At present, ELISA, Western blot, and ICT are frequently employed. The International Commission on Trichinellosis recommends ELISA as the primary test with Western blot as a confirmatory test, and these tests are also validated by WOAH. These tests are approved only for epidemiological surveillance and are not recommended for animal diagnosis for food safety purpose. This is because pigs can harbor the muscle larva before antibody reaches detectable levels, and seroconversion depends on the infective dose of the parasite.

ELISA

The antigen used in ELISA is the excretory-secretory (ES) component obtained from the stichocyte cells of live larvae of *T. spiralis* (TSL-1 antigen). This is a conserved antigen present in all species of *Trichinella*, and hence ELISA can detect antibodies to all 13 taxa of this parasite in pigs and other animals. Any of the samples of serum, whole blood, plasma, or tissue fluid collected before or after slaughter can be used. It has been found to be positive even when only one larva per 100 g of tissue is present. The IgG antibodies appear a minimum of 3 weeks after exposure, since *T. spiralis* larvae settle in pig muscle tissue approximately 17 days after infection.

Western Blot

The WB also uses the ES component of *T. spiralis* and is considered more sensitive and specific than ELISA. A specific triad of bands at 45, 49, and 53 kDa are considered positive for *Trichinella* infection.

Immunochromatographic Test

An ICT using larval and pre-adult *T. spiralis* excretory-secretory antigen has been described. It is based on Eu (III) nanoparticles (EuNPs) and uses a fluorescent probe. According to Wang and colleagues, this test detected anti-*Trichinella* IgG in pigs 4–5 days earlier than other existing tests using muscle larva antigens. The test had a turnaround time of 10 minutes and showed no cross-reactivity with other parasitic infections.

Taeniasis

The larva of the cestode belonging to Genus *Taenia* can cause a condition known as cysticercosis in animals. The important species are *Taenia solium, Taenia saginata*, *Taenia asiatica*, *Taenia multiceps*, and, rarely, *Taenia ovis* and *Taenia hydatigena*. All these species, except *T. hydatigena*, have zoonotic potentials. The definitive hosts include dogs, cats, humans, and wild carnivores in whom the parasite causes primary intestinal manifestations (intestinal taeniasis). The intermediate hosts include cattle, pigs, goats, sheep, and other ruminants. *T. saginata* primarily infects the bovine population, *T. solium* and *T. asiatica* are prevalent in pigs, and the other species are found in small ruminants. Most infections in animals caused by the adult and larval forms are asymptomatic. Acute *T. multiceps coenurosis* and *T. hydatigena* infections in sheep and goats are rare but may be fatal. Cysticercosis in animals leads to economic losses because the infected meat is not fit for consumption and must be discarded. *Taenia* infections have worldwide prevalence, and porcine cysticercosis is more common in regions with poor sanitation where pigs are free-ranging in the community.

Antigen detection: Antigen detection kits are useful for detecting infected carcasses in slaughterhouses and for individual diagnosis, but they suffer from low sensitivity due to low number of larvae in infected pigs and cattle. A commercial ELISA kit based on the B158/B60 antigen detection has been found to have poor sensitivity and specificity and a low positive predictive value of 35.2%. Notwithstanding the drawbacks, antigen ELISA has a role in epidemiological studies for exploring transmission dynamics in cattle and pigs and discovering point sources of epidemics or outbreaks.

Antibody detection: Tests for circulating antibodies have no role in diagnosis and can be used only for epidemiological studies. ELISA and electro-immune transfer blot (EITB) have been used for diagnosis of *T. solium* in pigs. EITB with a one-band cutoff had good sensitivity (89%) but poor specificity (48%). Increasing the cutoff to three bands increased the specificity but with a concomitant decrease in sensitivity.

Echinococcosis

Echinococcosis or hydatid disease is caused primarily by two species: *Echinococcus granulosus* and *Echinococcus multilocularis*. Other species of importance are *Echinococcus oligarthus*, *Echinococcus vogeli*, and *Echinococcus shiquicus*. *E. granulosus*, the causative agent of cystic echinococcosis, has worldwide prevalence, while *E. multilocularis*, causing alveolar echinococcosis, is more prevalent in the northern hemisphere. *E. vogeli* and *E. oligarthus* are mainly found in Central and South America, while *E. shiquicus* has been described in China. All the species, possibly with the exception of *E. shiquicus*, have zoonotic potentials.

Canines and felines including the wild carnivores are the definitive hosts in whom the condition is common. In dogs and cats, the parasite resides in the small intestine without causing any disease manifestations. A few cases of alveolar echinococcosis have been reported in dogs where these animals have accidentally served as intermediate host. Younger animals are more at risk of infection. Humans act as intermediate and dead-end host where cystic echinococcosis manifests as hydatid cysts in organs such as liver, lungs, kidney, and, rarely, brain. Alveolar echinococcosis in humans is a serious infection and potentially fatal. A wide range of animals act as intermediate or aberrant hosts. It includes rodents, cattle, sheep, goats, pigs, and horses. In these animals, cysts are found in internal organs, same as in humans.

Serodiagnosis in animals rely on both antigen and antibody detection in definitive or intermediate hosts.

Coproantigen Detection

Specific coproantigen(s) can be detected by antibody capture ELISA in dogs, cats, and other animals that are definitive hosts. It is particularly useful to detecting ongoing infection; it becomes positive within 5–10 days of exposure and becomes negative after successful treatment. Both polyclonal and monoclonal antibodies have been used. Purified IgG polyclonal antibodies raised in rabbits against crude somatic worm extracts or ES component from immature intestinal stages of *E. granulosus* have been used in ELISA with comparable results. Somatic preparations are easier to prepare but may be less specific compared with the ES component. Murine monoclonal antibody against somatic antigen of adult *E. multilocularis* has also been used for coproantigen detection. The specificity of the *E. granulosus* antigen test was high (96%) and sensitivity was comparable (92%) when the worm burden was high (>100) but came down to 30%–70% with worm burden <100. Nonetheless, antigen detection can be used for identifying infected dogs during control programs, and sensitivity is significantly better than antibody serology. Coproantigens can also detect prepatent infections in dogs.

Antibody Detection

E. granulosus protoscolex antigen preparation has been used in ELISA for antibody detection. It has variable sensitivity in different geographical locations (35%–73%) but has good specificity (70%–95%). The test becomes positive in 2 to 3 weeks. Later, an EM-2 was developed, but it too suffered from low sensitivity. Antibody detection by ELISA is useful in population-based epidemiological surveys but not very accurate in individual diagnosis in a host animal. It may help as a surveillance strategy in hydatid disease control programs.

In alveolar echinococcosis, ELISA based on native and recombinant antigens of the *E. multilocularis* metacestode such as Em2, Em2plus, CH-10, N3C, and the like has been used. To differentiate between alveolar and cystic echinococcosis, a Western blot test (Echinococcus Western Blot IgG; LDBIO Diagnostics, Lyon, France) detecting IgG against *E. multilocularis* whole larval antigens (Em16 and Em18) is available, accurately distinguishing the two conditions in 76% of cases.

Serodiagnosis in Intermediate Hosts

Diagnosis of echinococcosis is generally not indicated in farm animals such as cattle, sheep, or pigs, but population surveys are important, particularly in disease control programs. ELISA and other tests have been tried in sheep, but they have low sensitivity and poor specificity due to cross-reactivity with *Taenia* infections. Hydatid fluid antigen-based assays have sensitivity and specificity approaching 90%. At present, there is no acceptable sensitive and specific serological test available for livestock animals.

***Dirofilaria* (Heartworm) Infection**

Dirofilaria immitis and *Dirofilaria repens* are two filarial nematodes that cause heartworm disease in dogs and cats. Zoonotic infections in humans are also frequently observed; these parasites reside in the heart and the big vessels, and the adult female releases the larval form (microfilaria) in the bloodstream. The affected animals show pulmonary hypertension, arrythmias, and right heart failure. *D. immitis* in dogs is found globally, while *D. repens* is exclusive to the Old World.

ELISA and immunochromatographic tests have been used to detect antigens in canine populations and are widely employed in veterinary practice.

Antigen Test

The test detects a glycoprotein antigen produced by the mature female worm of *D. immitis* in its reproductive tract. Both ELISA and rapid methods are available to detect this antigen. It has high sensitivity (98%–100%) but can detect the infection only after 6 months when the female worm attains maturity and at least one or two females are present. The test becomes negative if only male worms are present. False-positive results have been noted with *D. repens*, *Angiostrongylus vasorum*, *Spirocerca lupi*, and *Acanthocheilonema dracunculoides* infections. If clinical suspicion is high but the antigen test is negative, heat treatment of the serum may be attempted to improve test results.

CONCLUSION

Immunodiagnostics for parasitic infections in animals has received limited attention, similar to the situation in human infections. This perspective is slowly changing with the development of commercial kits for use in animal infections and the repurposing of serological tests developed for human infections for diagnosis in animals. The later approach works well with zoonotic parasitic infections, which are substantial in number. Similar tests for exclusive animal infections are few (Table 16.1), and WOAH has been trying to bridge the gap through regional reference laboratories that undertake immunodiagnostic investigations for some of the neglected animal parasitic infections.

TABLE 16.1
Immunological Tests Used in Some Important Animal Parasitic Infections

Animal Infection	Parasite	Immunological Tests	Remarks
Babesiosis in bovines and canines	*Babesia bovis, Babesia bigemina, Babesia divergens, Babesia canis, Babesia vogeli, Babesia gibsoni*	ELISA IFAT	Used for bovine and canine babesiosis. Species identification due to cross-reactivity is a drawback
		ICT	Canine infections
Toxoplasmosis in felines	*Toxoplasma gondii*	ELISA, IFAT, Western Blot, ICT	Feline infections. IgM or high IgG indicates acute infections
Leishmaniasis in canines	*Leishmania infantum, Leishmania brazilensis, Leishmania tropica*	IFAT	Gold standard. Cross-reactivity with trypanosomes
		ELISA	Use of recombinant or chimeric protein antigens increases test sensitivity
		ICT	Recombinant fusion proteins rK39 and rK28 used as antigen
Sarcocystosis in cattle, pigs, sheep, canines, and felines	*Sarcocystis cruzi, Sarcocystis hirsuta, Sarcocystis hominis, Sarcocystis tenella, Sarcocystis caninum*, and others	IFAT, ELISA	Can be used for infections in cattle, goats, sheep, pigs, cats, and dogs
Nagana (trypanosomiasis) in cattle	*Trypanosoma congolense, Trypanosoma vivax, Trypanosoma brucei*	IFAT, ELISA	Only a presumptive diagnosis is possible in cattle, since antibodies persist for an extended period. Cross-reactivity with other trypanosomes is another problem

(*Continued*)

TABLE 16.1 (Continued)

Animal Infection	Parasite	Immunological Tests	Remarks
Surra in cattle, buffaloes, horses	*Trypanosoma evansi*	IFAT	
		ELISA	Universal antigen-based ELISA shows cross-reactivity with other trypanosomes. Can be used in all animals in various geographical locations
			Variant surface glycoprotein (VSG) antigen-based ELISA is highly specific
		Card agglutination test	Sensitivity varies depending on the animal species. Cross reactivity is also seen
		Immune trypanolysis test	Detects specific antibodies directed against variable antigen type RoTat 1.2. High specificity
Trichinellosis in pigs, wild carnivores	*Trichinella spiralis, Trichinella britovi, Trichinella nelsoni, Trichinella nativa*, and others	ELISA	TSL-1 antigen is used. Generic test and can detect antibodies against all 13 taxa found in pigs and other animals
		Western blot test	More sensitive and specific for *T. spiralis*
		ICT	Specific for *T. spiralis*
Taeniasis in pigs, cattle, sheep, goats	*Taenia solium, Taenia saginata, Taenia asiatica, Taenia multiceps, Taenia ovis, Taenia hydatigena*	ELISA for antigen detection	Low sensitivity and specificity. Useful for epidemiological studies
		EITB and ELISA for antibody detection	Limited role for diagnosis. Useful for epidemiological studies
Echinococcosis in canines and felines, wild carnivores	*Echinococcus granulosus, Echinococcus multilocularis, Echinococcus.vogeli, Echinococcus oligarthus, Echinococcus shiquicus*	ELISA for antigen detection	In fecal samples in dogs and cats. High specificity and sensitivity
		ELISA for antibody detection	High specificity but variable specificity. More useful for population survey both in definitive hosts and intermediate hosts (sheep, cattle, and pigs)
Dirofilariasis (heartworm infection) in dogs and cats	*Dirofilaria immitis, Dirofilaria repens*	ELISA and ICT for antigen detection	High sensitivity but becomes positive only after 6 months of exposure. Positive only if female worm is present

FURTHER READINGS

Bajyana OE, Hamers R. A card agglutination test (CATT) for veterinary use based on an early VAT RoTat 1–2 of *Trypanosoma evansi*. *Ann Soc Belg Med Trop.* 1988;68:233–240.

Bruschi F, Gómez-Morales MA, Hill DE. International Commission on Trichinellosis: Recommendations on the use of serological tests for the detection of *Trichinella* infection in animals and humans. *Food Waterborne Parasitol.* 2019;14:e00032.

Desquesnes M, Bossard G, Thevenon S, Patrel D, Ravel S, Pavlovic D et al. Development and application of an antibody-ELISA to follow up a *Trypanosoma evansi* outbreak in a dromedary camel herd in France. *Vet. Parasitol.* 2009;162:214–220.

Parija SC, Chaudhury A, editors. *A Textbook of Parasitic Zoonoses*. Berlin: Springer International; 2022.

Raether W, Hänel H. Epidemiology, clinical manifestations and diagnosis of zoonotic cestode infections: An update. *Parasitol Res.* 2003;91:412–438.

Rodriguez S, Wilkins P, Dorny P. Immunological and molecular diagnosis of cysticercosis. *Pathog Glob Health.* 2012;106:286–298.

Torgerson PR, Deplazes P. Echinococcosis: Diagnosis and diagnostic interpretation in population studies. *Trends Parasitol.* 2009;25:164–170.

Wang X, Tang B, Zhao Y, et al. Development of a rapid and sensitive immunochromatographic strip based on EuNPs-ES fluorescent probe for the detection of early *Trichinella spiralis*-specific IgG antibody in pigs. *Vet Res.* 2021:52:85.

World Organisation for Animal Health. Codes and Manuals. https://www.woah.org/en/what-we-do/standards/codes-and-manuals/

Part III

Recent Trends in Immunodiagnostics of Parasitic Infections

17 Rapid Point-of-Care Tests

INTRODUCTION

Most of the currently available immunological tests used for parasitic and other infections depend on trained operators, require instrumentations, and are time consuming. This has prevented their use in the field scenario and at patients' bedside, since a well-equipped laboratory is a prerequisite for these tests. The first breakthrough in the development of a rapid test that can be performed by untrained or minimally trained individuals first appeared in the 1970s. The test was developed for pregnancy testing, and it rapidly gained widespread adoption due to the simplicity of the testing process. This has prompted the commercial development of several such tests for both infectious and noninfectious diseases for antigen or antibody detection. Dot ELISA, Dot ELISA dipsticks, and lateral flow immunochromatographic tests (ICT) are the three available formats of rapid tests with a turnaround time ranging from 15 to 120 minutes. The rapid tests are widely used for several parasitic diseases, with malaria detection kits heading the list. This is also used for other parasitic infections such as amoebiasis, giardiasis, cryptosporidiosis, leishmaniasis, lymphatic filariasis, onchocerciasis, schistosomiasis, and others.

CRITERIA FOR POINT-OF-CARE TESTS

Point-of-care (POC) tests are designed to be performed close to the patient without any need for sample transportation to the laboratory. Thus, they are ideal for field conditions for screening of infected individuals and for epidemiological surveillance. The World Health Organization, Centers for Disease Control and Prevention, and the Foundation of Innovative Diagnostics came together and adopted the ASSURED criteria for POC tests in 2015, a term already coined in 2003. ASSURED stands for **A**ffordable, **S**ensitive, **S**pecific, **U**ser-friendly, **R**apid and robust, **E**quipment free, and **D**eliverable to end-users. In recent years, two other components have been prefixed: **R**eal-time connectivity and **E**ase of collection, to create the acronym REASSURED (Table 17.1). This forms the yardstick for all POC tests and is particularly useful for resource-challenged areas of the world.

PRINCIPLE OF THE TESTS

All rapid POC immunological tests are based on either sandwich assays or competitive assays. Another method that has gained popularity in recent years is the multiplex assay.

Sandwich Assays

In the antigen detection assay, enzymes-, nanoparticles-, or fluorescence dye–coated antibody (labeled antibody) is immobilized at a conjugate pad. This antibody can be mobilized when the buffer solution is added. A capture antibody against target antigen is immobilized over the test line. A secondary antibody against labeled antibody is immobilized at the control line. Once the sample is applied to the designated area, it migrates along the strip. At the conjugate pad area, the analyte is captured by the labeled antibody to form a complex. This complex then migrates along the membrane by capillary force. At the test line area, this complex is again captured by a primary antibody to the analyte molecule. As a result, the analyte gets sandwiched between the primary antibody and the labeled antibody. The excess labeled antibody gets captured by the secondary antibody at the control line. The intensity of the color developed at the test line indicates the relative concentration of the analyte. The color developed at the control line ensures satisfactory functioning of the test. The color can be visualized or detected by a strip reader. Figure 17.1 shows the working of the sandwich assay.

DOI: 10.1201/9781003644224-22

TABLE 17.1
REASSURED Criteria

Acronym	Criterion	Interpretation
R	Real-time connectivity	Test can be connected and powered by mobile phones or reader and to provide test results or data directly to decision-makers
E	Ease of specimen collection	Preferably by noninvasive technique or minimum invasion (e.g., fingerprick blood sample)
A	Affordable	Economical both to the subject and the health system
S	Sensitive	No false negatives
S	Specific	No false positives
U	User friendly	Easy to perform by the operator with minimum training
R	Rapid and robust	Quick turnaround time (15–120 min). Hardy enough to minimize the need for specialized transport and storage
E	Equipment free or simple and environment friendly	No need for any special equipment or use of a simple reading device operated by battery or solar power. Test components can be safely disposed of without pollution hazards
D	Deliverable to end user	Can be accessed easily by those who need them most

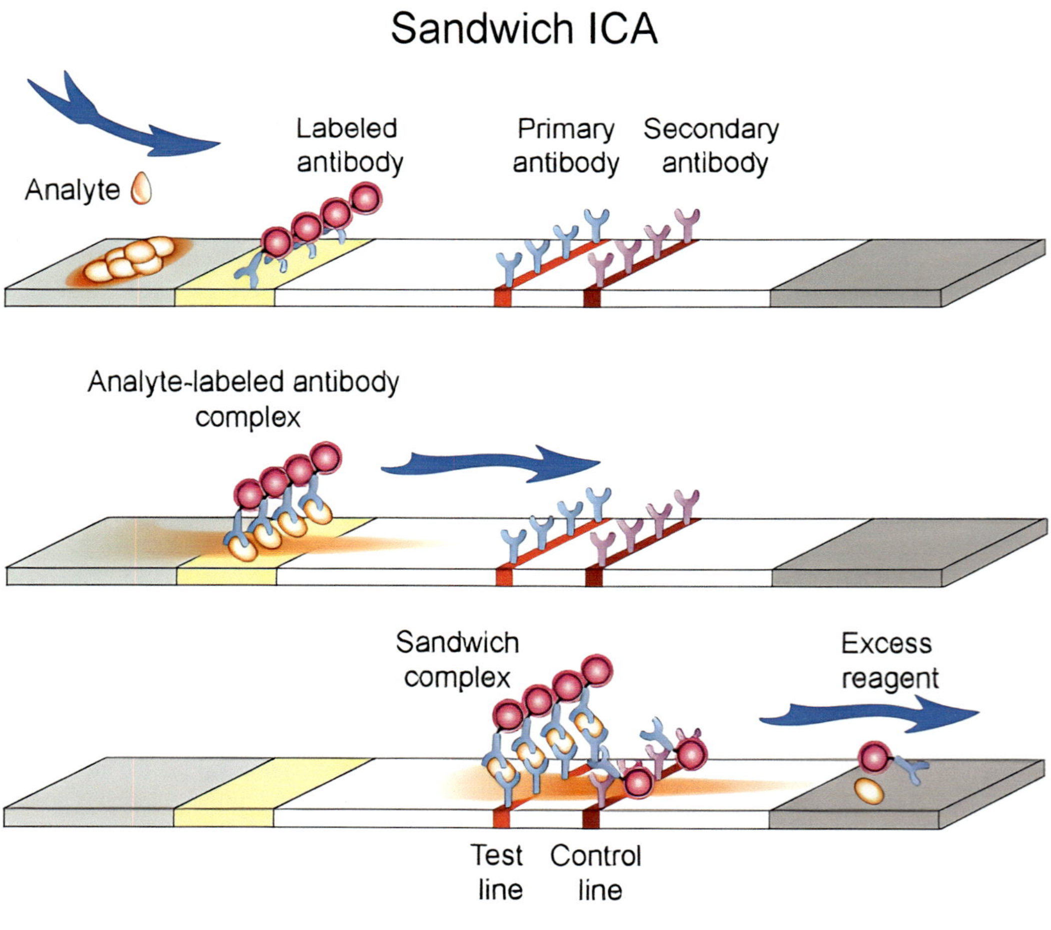

FIGURE 17.1 Principle of the sandwich assay.

(Photo credit: Creative Diagnostics.).

Competitive Assay

In this format, the sample application pad contains prefixed labeled antibody, and when the target analyte is applied, the sample with the hydrated labeled antibody starts migrating. The test line already contains the analyte in question, and it binds specifically to the labeled conjugate. When the liquid sample reaches the test line, the pre-immobilized antigen binds to the labeled conjugate if the same antigen is absent in the sample. In effect, the antigen in the sample solution and the one immobilized at the test line will compete to bind with the labeled conjugate. The control line contains secondary antibody that can bind with the labeled antibody. The absence of any color at the test line indicates the presence of antigen in the sample and hence a positive result. In case of a negative test, color develops both at the test line and the control line (Figure 17.2).

Multiplex Assay

Another method that has become popular in recent times is the multiplex assay. In this format, several test analyses can be detected simultaneously in a single test strip by increasing the relative length of the strip and increasing the test lines. Such a format is available commercially for *Giardia, Entamoeba*, and *Cryptosporidium* antigen detection in stool and the dual antigen detection of HRP2 and pLDH for detection of *Plasmodium falciparum* and other *Plasmodium* species using the same strip.

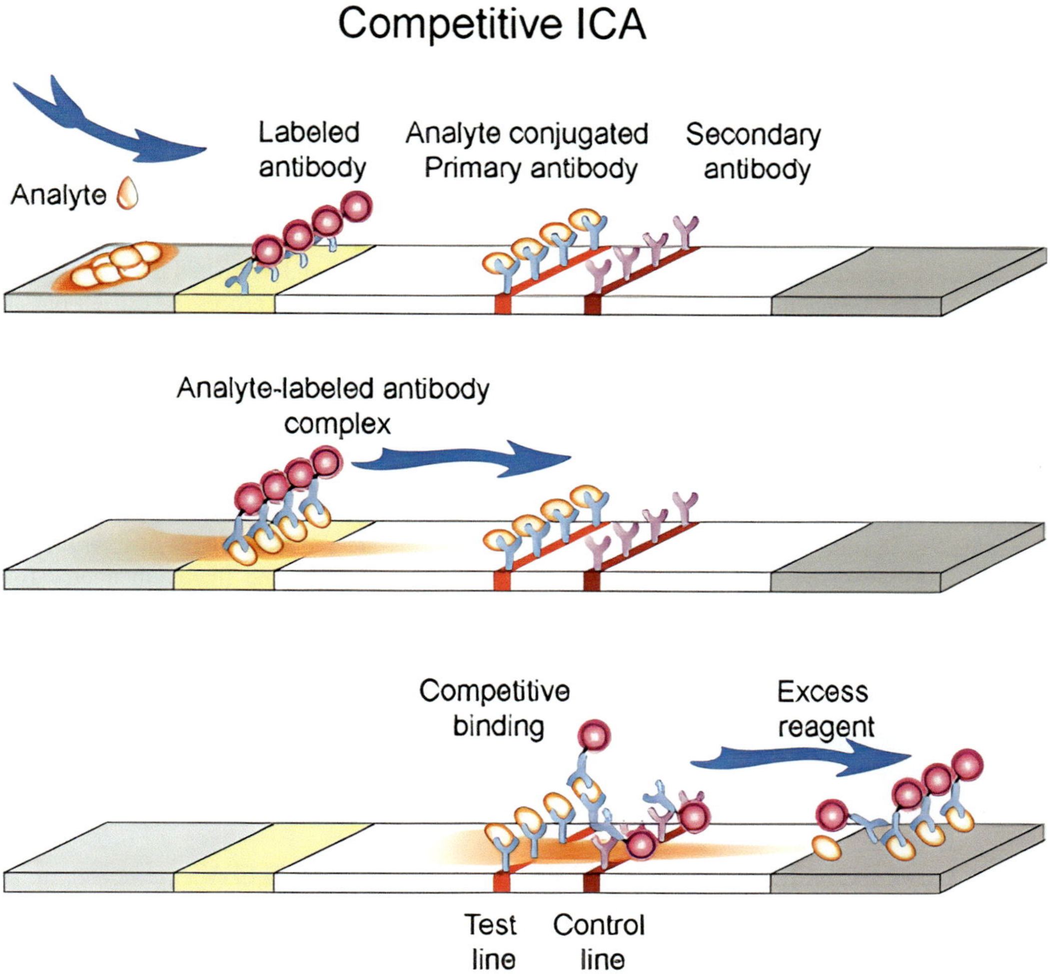

FIGURE 17.2 Principle of the competitive assay.

(Photo credit: Creative Diagnostics.).

Types of Rapid Tests

Three formats of rapid tests are available and produce comparable results.

Lateral Flow Immunochromatographic Assay (LFA)

This test is easy to use, economical, robust, can be of utility in remote areas for field studies, and is ideal for tropical and subtropical areas, since these tests work well at temperatures as high as 40°C. The interpretation can be done by visual inspection, although for improving the sensitivity, spectroscopic, fluorescent, or electrochemical readings can be taken using simple devices.

A chromatographic principle is used in this assay to separate various test components either by longitudinal or vertical flow through carrier membranes. The detector immunogens are already immobilized on the carrier over which the test sample flows, resulting in immunofixation of the analyte. The basic working of LFA is outlined in the sandwich assay above.

The LFA has been found to be of higher sensitivity compared with ELISA and IFAT. False-negative results can occur if the parasite load is low or if there is a very high level of the analyte leading to pre- or post-zone effects. Although LFA has been studied for various parasitic infections, the approved kits are available commercially for the following parasitic disease: malaria, visceral leishmaniasis, giardiasis, amoebiasis, cryptosporidiosis, filariasis, and schistosomiasis.

Dot-ELISA

This test belongs to the format of sandwich ELISA where the antigen is captured between the primary and secondary antibodies.

The basic principle is similar to the conventional ELISA except that nitrocellulose membrane is used instead of solid polystyrene wells as the assay zone. The appropriately labeled antibody is applied to a small circular spot on the membrane and, after adding a blocking solution, is incubated for 15 minutes. The test sample is then added and allowed to react for 30 minutes, followed by addition of conjugated antibody. The enzyme substrate is added next and incubated for 30 minutes, resulting in development of a colored spot in a positive test. The whole process is carried out at room temperature. It requires lesser amounts of sample as well as the primary and secondary antibodies compared with conventional ELISA, with similar sensitivity and specificity.

Dot-ELISA Dipsticks

The dipsticks are prepared before field visits using nitrocellulose strips impregnated with the appropriate antigen or antibody at specific areas to form test and control lines. These strips are then glued to acetate strips. In field conditions, these strips are incubated with test serum for 7 minutes, followed by conjugated antibody for another 7 minutes, and finally incubated for 2 minutes with the enzyme substrate. Washing steps need to be done in between each step. Development of colored bands at test and control lines signifies a positive test. The nitrocellulose membrane can also act as a serum separator; whole blood can be applied, and the serum separates due to capillary action and moves to the test area. Such tests are available commercially for malaria parasites. The LFA described above is a modification of Dot-ELISA dipsticks to improve the sensitivity of the tests.

IMPORTANT CONSIDERATIONS IN DESIGNING LATERAL FLOW IMMUNOCHROMATOGRAPHIC POC TESTS

The important components to be considered while designing an in-house lateral flow test include the membranes, the pad material (Figure 17.3), the capture antibodies and antigen, and the detector reagents.

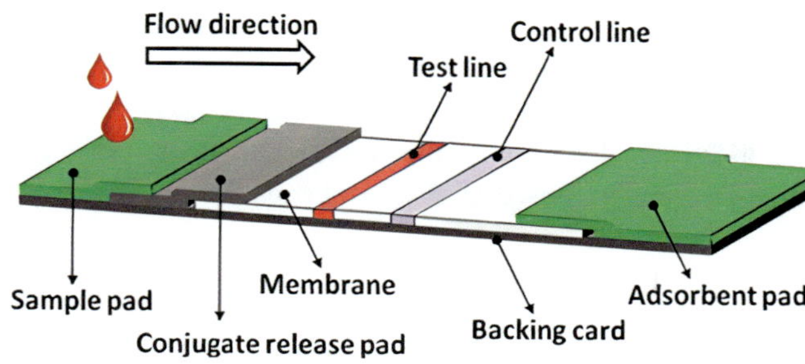

FIGURE 17.3 A diagram of a lateral flow device.

(From Koczula KM, Gallotta A. Lateral flow assays. *Essays Biochem*. 2016;60:111–120. Reproduced with permission.)

The Membrane Component

The membrane plays a crucial role in the test performance because of its properties that affect the capillary flow of the test reagents and analytes. This in turn affects the reagent deposition, consistency of the test lines, and ultimately the sensitivity and specificity of the test. The test and control lines are drawn on this component.

Various polymers have been used, but the nitrocellulose membrane is the preferred one for several reasons. It can capture reagents because of its excellent protein-binding capacity. Nitrocellulose membranes are neutral without any acidic proteins, binding protein by electrostatic bonding. When applying capture reagents to the membrane, agents like Tween-20 or TritonX-100 should be avoided because they interfere with the protein–nitrocellulose molecular interface. The thickness of the membrane is also important, since an increase in thickness affects the width of the test and control lines. Thinner membranes tend to spread the fluids over a larger area compared with thicker ones because of less depth available for penetration. Larger diffusion of reagents results in a decrease in the intensity of the color bands and less sensitivity of the test.

Pad Materials

These are porous materials used for the sample, conjugate, and absorbent pads. While filter papers are commonly used for sample and absorbent pads, glass fiber filters are used for the conjugate pad.

Sample Pad

The sample pad performs the important function of even distribution of the sample over the conjugate pad and thus helps transport the sample to other parts of the system. It also controls the rate of entry of the fluid into the conjugate pad to prevent flooding. Cellulose fiber filters and woven mesh are two types of material used in a sample pad, of which the former is most commonly used. Fiber filters are thick, have a larger bed volume, are cheap, and can be loaded with a large variety of reagents of differing pH and ionic strengths.

Conjugate Pad

Antibodies mainly labeled with colloidal gold nanoparticles are dispensed here. When the conjugate pad comes in contact with the moving fluid (sample), it should be able to release the labeled conjugate rapidly. The conjugate pad should have low nonspecific binding property, have consistent flow characteristics and bed volume, and be free of particle that can clog the membrane. The materials commonly used in the manufacture of conjugate pad are nonwoven filters made up of compressed cellulose, glass, or plastic (surface modified polyester or polypropylene) fibers. Of these, cellulose filters have the highest hold-up volume (>50 µL/cm). All have low nonspecific binding property.

Absorbent Pad

It is kept at the distal end of the device and acts as a sink. It helps increase the total volume of the sample entering the system. The excess volume effectively washes away unbound detector molecules from the test and control lines and decreases the background noise, thus enhancing sensitivity. Cellulose filters are usually the material used in absorbent pads.

Capture Antibodies

The antibodies should be highly purified and carefully designed. Affinity and specificity are two key aspects of the antibody that determine the efficacy of LFA. Affinity is important because the antigen, antibody conjugate, and the coated antibody react for only a few seconds. To this end, fast k_{on} and slow k_{off} rates are necessary. Specificity of the antibody is also essential so as not to detect similar proteins. The antibodies are coated on the strip at a high concentration (1–3 µg per linear cm) over a small bed volume of 0.1 mm^3, which is 25–100 times greater than those coated in ELISA plate wells.

These antibodies can be prepared in the laboratory or purchased commercially. Either monoclonal or polyclonal antibodies can be used. Monoclonal antibodies offer excellent specificity at the cost of sensitivity. There may, however, be lot-to-lot variation in the concentration and purity of the monoclonal antibodies, which necessitates screening of these antibodies before use, which is expensive. Polyclonal antibodies, on the other hand, are cheaper and easier to produce in animals such as rabbit, goat, chicken, and others. They offer higher sensitivity because of the presence of a mixture of antibodies that can recognize different epitopes of the antigen, increasing the signal strength. Here, too, there may be variation among different lots produced.

The control line in the LFA indicates proper functioning of the test, and conjugated secondary antibodies are used to create the control line. Anti-chicken secondary antibodies are commonly used, which bind to chicken IgG. Chicken IgG shows minimum nonspecific interaction because it differs widely from mammalian IgG in homology.

Detector Reagents

Various materials have been used as detector or reporter molecules, of which the most common ones are gold nanoparticles, latex beads, and fluorescent dyes (Figure 17.4).

Results / Labels	Quantitative	Semi-quantitative	Qualitative
Color labels e.g., gold nanoparticles, colored latex	Optical strip readers or camera (with imaging software) for measurement of the intensity of colors produced at test and control line	Visual inspection of ladder bar, where the number of colored lines is an indication of analyte concentration	Visual inspection of line colors
Fluorescent labels e.g., quantum dots, ruthenium complexes	Fluorescent strip reader, recording fluorescence intensity		
Other labels e.g., paramagnetic labels, enzyme labels, carbon nanoparticles	Magnetic strip readers; Electrochemical detectors; Chemiluminescence readers		

FIGURE 17.4 Various detector systems used in lateral flow assays.

(From Koczula KM, Gallotta A. Lateral flow assays. *Essays Biochem.* 2016;60:111–120. Reproduced with permission.)

Gold Nanoparticles

These are by far the most commonly used because of the intense signal gold produces and the high stability in both liquid and dried forms. The conjugation process is also easy to perform, and it is available commercially of uniform quality. The color intensity is due to the plasmon absorbance and light scattering by the electronic shell of gold atoms. Particles of 40nm are widely used for LFA. Conjugation with antibody occurs due to electrostatic and hydrophobic interactions. Antibody and gold particles are mixed in a low-ionic-strength buffer and then blocked with casein, albumin, or polyols.

Latex Particles

This is another cheap alternative as reporter material. Particles can be loaded with a dye, a fluorophore, or paramagnetic particles. Latex of different colors can be used in multiplex assays. The particle size ranges from 100 to 300 μm, and this large size lowers the sensitivity of the assay because of less dense packing at the test and control lines.

Fluorescent Dyes

Fluorescein and cyanine dyes are used because of their small size (<1 nm), which allows them to be released more easily from the conjugate pad. They can be conjugated covalently with the primary amines and sulfhydryl groups of proteins. However, the readings need to be taken with specialized readers.

Buffers

The buffers used to prepare the capture reagents need to be optimized for proper functioning of the immunochromatographic device. The buffer so chosen must not interfere with protein absorption, reagent activity, and membrane properties. Optimum pH should be maintained at 7.0–7.5. Various buffers can be used, including phosphate, borate, carbonate, and TRIS. For capture reagents that are stable at pH 4.0–6.0, ammonium acetate buffer can be used. The key elements while choosing a buffer include:

a. Very low molarity (<10 mM).
b. NaCl free.
c. Free from any detergent or surfactants (Tween-20, Triton X-100). However ionic detergents like SDS (<0.05% w/v) may be used if needed for stabilization.
d. No other additives present.

Blocking Agent for Membrane

The use of a blocking agent in the lateral flow assay system remains optional, since there are commercial kits where blocking agents have not been used and they work satisfactorily. The other protein fractions present in the sample or the blocking agent added to the sample pad are supposed to be sufficient for prevention of nonspecific binding of analyte or detector agents to the nitrocellulose membrane. However, if a blocking agent is used, several options are available. These include bovine serum albumin (BSA) (1%–2% w/v), polyvinyl alcohol (PVA) (0.1%–1% w/v), gelatin (0.1%–0.5% w/v), and others applied for 1–15 minutes depending on the agent. After application of the blocking agent to the membrane, it is essential to remove excess blocking agent with 5 mM Na_2HPO_4 (pH 7.5) with added 0.01% w/v SDS applied for 5–15 minutes.

Design and Optimization of *Toxoplasma gondii* Antibody Detection Rapid Test

Here we describe an experimental test kit for rapid diagnosis of *Toxoplasma gondii* infection as reported by Wang et al. Another POC test for detection of *Trichinella spiralis* antibody has been described by Thanchomnang et al.

Reagents and Equipment Required

Reagents: NI-IDA agarose beads; SDS page; Tris base; glycine; methanol; Tween-20; nonfat dry milk; horseradish peroxidase; NaCl; HNO_3; chlorinated organopolysiloxane in heptane (Sigmacote); colloidal gold nanoparticles; bovine serum albumin; PBS; sucrose; sodium azide.

Equipment and labware: dialyzer; affinity chromatography instrumentation; electrophoresis chamber with power pack; membrane blotter system; chemiluminescence detection system; PVC plate; molded plastic casings (cartridges); micropipettes; glassware; PVDF membrane; nitrocellulose membrane; glass fiber conjugate pad; cellulose fiber sample pad.

Others: MIC2 and MIC3 gene constructs of *T. gondii*; *Escherichia coli*; pET-32a plasmid vector; *E. coli* Transetta (DE3) strain; rabbit anti–*T. gondii* serum; goat anti-rabbit IgG; Protein A.

Design and Optimization of the Test

The design of the *Toxoplasma gondii* antibody detection rapid test can be divided into the following steps.

Antigen Preparation

The antigen used for the test is the recombinant protein combination of Microneme protein 2 and 3 (MIC2, MIC3). MIC2 is part of parasite adhesins molecules, while MIC3 shows strong immunogenicity and is present in all stages of the parasite: bradyzoites, tachyzoites, and sporozoites.

- The MIC2 and MIC3 gene constructs of *T. gondii* are designed based on the sequence of proteins available in the NCBI gene bank.
- Using ExPASy-Prot Scale, two sequences are selected from regions based on high hydrophilic fusion content, 267–345 aa sequence for MIC2 and 234–307 aa for MIC3, and the two proteins are fused using GSGSG flexible peptide.
- These amino acid sequences in the fusion protein are converted to nucleotide sequence, and the codon is optimized based on *Escherichia coli* preferences.
- The coding sequences are biosynthesized and cloned in the pET-32a plasmid vector for expression.
- The recombinant protein is then generated by transfection into *E. coli* Transetta (DE3) of pET-32a-MIC2-MIC3-His complex. His-tagged fused proteins are purified by affinity chromatography using Ni-IDA agarose beads.
- The purified protein fraction in the elution buffer was freed from imidazole by dialysis and exchanged into phosphate buffered saline (pH 7.4).

Antigen Confirmation

To ascertain whether the appropriate antigen has been generated, the following blotting procedure is done.

Purified protein, 10 μg is separated by SDS-PAGE and then transferred to polyvinylidene difluoride membrane using wet transfer system equipment (blotter) at 100 V for 1 hour in a transfer buffer made of 25 mM Tris, 192 mM glycine, and 20% methanol (pH 8.3).

Antigen blocking is done with 5% nonfat dry milk in Tris-buffered saline with 0.1% Tween-20 (TBST) for 1 hour at room temperature.

Rabbit anti–*T. gondii* serum is diluted to 1:100 in 5% nonfat dry milk in TBST, and the membrane is incubated with the serum overnight at 4°C.

Next day the membrane is washed thrice for 5 minutes each and incubated with horseradish peroxidase conjugated with goat anti-rabbit IgG, 1:5,000 (secondary antibody, available commercially) for 1 hour at room temperature.

After additional wash, protein bands are visualized using an enhanced chemiluminescence detection system.

Gold-Conjugated Protein A Preparation

The glassware used for gold conjugate preparation are first cleaned with aqua regia HNO_3/HCl (v/3v) and then treated with Sigmacote. This is a siliconizing reagent (chlorinated organopolysiloxane in heptane) for glass and other surfaces which forms a thin microscopic layer on the glass and prevents surface adsorption of various proteins.

Colloidal gold nanoparticles can be obtained from commercial source or can be prepared as follows.

- A 0.01% aqueous solution of hydrogen tetrachloroaurate ($HAuCl_4$) is prepared and 100 ml of the solution was brought to boiling.
- To the solution, 2.5 ml of 1% trisodium citrate solution is added with constant stirring.
- Further boiling for 10 more minutes resulted in color change from blue to dark red, indicating colloid gold formation.
- The heating is stopped, and the suspension stirred for further 5 minutes to stabilize.
- 0.01% sodium azide is added, and the suspension is stored at 4°C in a dark container.

Conjugation of Protein A with Gold

- Protein A solution (1.0 mg/mL) is gently mixed into 100 mL of colloidal gold suspension.
- The mixture is stirred and 1% (w/v) of BSA is added to block the conjugate.
- The whole preparation is centrifuged, and the conjugate pellet is resuspended in 0.01mM of PBS containing the following: 1% (w/v) BSA, 0.3% (v/v) Tween-20, 0.9% (w/v) NaCl, and 0.05% (w/v) sodium azide.
- The preparation is stored at 4°C.

Immunochromatographic Strip Preparation

- Nitrocellulose membrane
  - The test line and the control line are made on the membrane. Diluted fusion proteins MIC2-MIC3 and anti–Protein A IgG are painted on the membrane in 1 μL/cm volume to form the test and the control lines, respectively.
  - Subsequently the membrane is treated with 0.02 MPBS containing 1.0% BSA and dried at 37°C for 2 hours.
  - For the absorbent pad, pure cellulose is used.
- Sample and conjugate pads
  - The sample pad (cellulose fiber) is prepared by saturating it with 0.02 M PBS solution (pH 8.5) containing 1.5% (w/v) of BSA and 0.2% Tween-20.
  - It is then dried at 37°C for 2 hours.
  - For the conjugate pad, the colloidal gold suspension is first diluted 1:5 (v/v) with 0.02 M PBS (pH 8.5) containing 1.5% (w/v) BSA and 5% (w/v) sucrose. The glass fiber conjugate pad is first treated with 0.2% Tween-20 for 12 hours and then dried at 37°C. The gold conjugate is the added and then dried at 37°C for 2 hours.
- All the test strip components—nitrocellulose membrane, absorbent pad, sample pad, and conjugate pad—are assembled on a PVC plate, and the assembled strips are cut into strips 6 mm in width each. The strips are kept in a desiccator at 4°C for test use.

- Strips can also be placed into molded plastic casings to ensure that the sample is applied in the correct area and that proper flow is achieved. A test window with appropriate markings is also included in the plastic cartridge to facilitate correct result readout.

Test Procedure

- Test serum sample is diluted at 1:4, and 100 μL of the sample is dispensed on the sample pad.
- Reading is taken after 10 minutes.

Result and interpretation: Presence of a pink/red line both at the control and the test signifies a positive test. If there is no color band at the control line, the test is considered invalid.

Test performance: The LFA is compared to known positive and negative samples as determined by a commercially available ELISA. The coincidence rate of the test was 94.6%, while the sensitivity and specificity were 100% and 85.7%, respectively. The test strips performed well even after storage for 3 months at 24°C.

Quality assurance: The internal reagent control line in each strip must give a colored line to validate the assay. In addition, a known positive and negative quality control serum must be run with each batch of patient testing and for each new lot if using commercial kits.

Troubleshooting in LFA tests: There may be various issues with the LFA tests either obtained commercially or prepared in the laboratory. Table 17.2 summarizes the most important problems encountered with the possible solutions.

TABLE 17.2
Common Troubleshooting Guide

Problem	Solution
False-positive signal	Modify buffer in conjugate pad/solution (e.g., pH, salt concentration, or surfactant concentration) Change conjugated protein Use blocking agent
False-negative signal	Modify buffer in conjugate pad/solution (e.g., pH, salt concentration, or surfactant concentration) Change conjugated protein Use membrane with smaller pore size Increase sample volume
Uneven lines or dots	Use membranes with different pore size Reduce dispensing volume of reagent Increase protein concentration of reagent Check dispensing buffer composition Check dispensing process
Uneven fronts of migrating sample	Check membrane shelf life Use membrane with different/more surfactant Check whether the relative humidity is very low Increase surfactant concentration in conjugate pad
Background on nitrocellulose membrane	Pretreat the sample pad with surfactants like Tween-20 or Triton-100 Change the orientation between membrane and absorbent pad Use membrane with higher wick thickness
Conjugated particle not released from the pad	Increase the conjugate blocking Use monoclonal antibody instead of polyclonal
Low stability	Trehalose or sucrose to be used as stabilizers Airtight pouches with activated silica to be used for device storage
Captured antibody spreads out while coating	Methanol or ethanol (1%–3%) to be used in capture antibody buffer

TABLE 17.3
Experimental and Commercial Rapid Point-of-Care Tests for Parasitic Diagnosis

	Protozoa		Nematode		Cestode		Trematode	
	Antigen	**Antibody**	**Antigen**	**Antibody**	**Antigen**	**Antibody**	**Antigen**	**Antibody**
Commercially available test	*Plasmodium falciparum* *Plasmodium vivax* Pan malaria parasites *Entamoeba histolytica* *Giardia intestinalis* *Cryptosporidium* spp. *Trichomonas vaginalis*	*Toxoplasma gondii* *Leishmania* spp. *Trypanosoma cruzi* *Trypanosoma brucei gambiense*	*Wuchereria bancrofti* *Onchocerca volvulus*	Nil	Nil	*Echinococcus granulosus*	*Schistosoma mansoni* *Schistosoma haematobium*, *Schistosoma japonicum*	Nil
Research/ experimental tests	*Leishmania* spp. *Acanthamoeba* spp *Babesia* spp. (for animals)	*Brugia* spp.	*Angiostrongylus cantonensis*	*Trichinella* spp. *Capillaria philippinensis* *Gnathostoma* spp. *Toxocara* spp.	NIL	*Echinococccus multilocularis* *Taenia solium*	Nil	*Fasciola hepatica* *Paragonimus* spp.

Rapid Point-of-Care Tests for Parasitic Infections

Several rapid tests based mainly on LFA have been developed experimentally by various workers, and some such assays are available commercially. Most of these tests have a turnaround time of 5–20 minutes, though a few may require 1–2 hours. These have become invaluable tools not only in health care facilities but also in the field situation. Table 17.3 summarizes the various rapid tests available commercially or developed experimentally. The number of tests available for malaria parasites far outweighs the tests available for other protozoan or helminthic infections. For the full list of commercially available POC tests for parasites, please refer to Appendix B.

CONCLUSION

Rapid POC tests (including rapid DNA-based technologies) are here to stay, and their popularity is on the increase because of faster results, ease of use, and affordability. Various newer strategies are being experimented with to increase assay sensitivity. These include the use of silver enhancement technology and combining horseradish peroxidase with gold particles, magnetic particles, and others, which can enhance detection to 0.1 ng/ml of the analyte. Signal amplification can be increased 1,000-fold using thermal contrast, laser, or light-emitting diode. Multiplex assays to detect two or three analytes simultaneously are also being used widely. Although antigen detection test is ideal to pinpoint active infection, antibody detection may not be helpful in many instances, since these tests cannot differentiate between recent and past infection and cannot be relied on when used in endemic areas. Notwithstanding the limitations of rapid tests, it is expected that more such tests will be introduced in the near future, and that many research or experimental assays will reach commercial production and become available on a larger scale for wider application.

FURTHER READINGS

Abaza SM. Immunochromatographic assays in diagnosis of parasitic diseases. *Parasitol Union J.* 2008;1:1–13.

Koczula KM, Gallotta A. Lateral flow assays. *Essays Biochem.* 2016;60:111–120.

Momčilović S, Cantacessi C, Arsić-Arsenijević V, Otranto D, Tasić-Otašević S. Rapid diagnosis of parasitic diseases: current scenario and future needs. *Clin Microbiol Infect.* 2019;25:290–309.

Rogers MJ, McManus DP, Muhi S, Gordon CA. Membrane technology for rapid point-of-care diagnostics for parasitic neglected tropical diseases. *Clin Microbiol Rev.* 2021;34:e00329–20.

Sukumaran A, Thomas T, Thomas R, Thomas RE, Paul JK, Vasudevan DM. Development and troubleshooting in lateral flow immunochromatography assays. *Indian J Clin Biochem.* 2021;36:208–212.

Thanchomnang T, Sadaow L, Sanpool O, Intapan PM, Rodpai R, Boonroumkaew P, et al. Development of an immunochromatographic point-of-care test fordetection of IgG antibody in serodiagnosis of human trichinellosis. *Int J Infect Dis.* 2021;111:148–153.

Wang J, Zhao Y, Qiu J, Li J, Zhou R, Ma X, et al. Development of an immunochromatographic test with recombinant MIC2-MIC3 fusion protein for serological detection of toxoplasma gondii. *VetSci.* 2025;12:509.

World Health Organization. *Laboratory and Point-of-Care Diagnostic Testing for Sexu+ally Transmitted Infections, Including HIV.* World Health Organization; 2023. https://iris.who.int/handle/10665/374252

18 Biomarkers and Evolving Molecular Immunodiagnostic Tests

INTRODUCTION

There has been tremendous technological advancement in the last two to three decades, opening new avenues for detection and diagnosis of infectious diseases. The expanding field of proteomics is opening with peptides unique for a given species or genus of a parasite, paving the way for design of highly specific immunoassays. Researchers in the physical sciences are working toward the development of novel platforms such as microfluidic technology and several electrochemistry-based sensor systems such as field-effect transistor, terahertz-wave-based methods, and optical-based methods such as surface-enhanced Raman scattering, which can tremendously improve the assay sensitivity and ease of operation as a lab-in-a-chip.

Apart from these new technologies, existing systems such as flow cytometry are being increasingly evaluated for the detection of protozoan parasites with high accuracy. Two well-tested existing systems, the enzyme-linked immunosorbent assay (ELISA) and polymerase chain reaction (PCR), have been integrated in the PCR-ELISA method with simultaneous enhancement of specificity and sensitivity. It is expected that many of these research methodologies will find broader applications in the near future, potentially leading to the phasing-out of some existing assay methodologies, as has occurred with the complement fixation test.

NOVEL IMMUNODIAGNOSTICS METHODS

Biomarkers

Biomarkers have been defined by the National Institutes of Health Biomarkers Definitions Working Group as "a characteristic that is objectively measured and evaluated as an indicator of normal biological processes, pathogenic processes, or pharmacologic responses to a therapeutic intervention." This definition has a broad connotation and encompasses all aspects from pulse rate of a patient to complex laboratory tests to detect a specific molecule. When used in the context of laboratory medicine, these markers may be directly related to the pathogen or disease process itself. In addition, it may be also an indirect method due to body's reaction to the disease, most commonly as a result of cellular or humoral immune response.

In immunodiagnostics, the biomarkers can be equated with antigens or some other molecule produced either by the pathogen or the individual, which can be detected by immunological methods. The earliest pathogen-produced biomarkers that came into wide use are those for malaria parasite, which include histidine-rich protein II, parasite lactate dehydrogenase, and aldolase. In recent years, proteomic studies have revealed various peptides produced by a parasite. Their characterization and functional analysis are elucidating exotic peptides of the pathogen useful for design of immunodiagnostics. Apart from malaria parasite, some other blood- and tissue-dwelling protozoan and helminth parasites have been analyzed fully or partially, even though the field as a whole remains largely an experimental arena. In his chapter, some of the biomarkers that have been explored in recent years are described.

DOI: 10.1201/9781003644224-23

Malaria Parasites

Plasmodium falciparum histidine-rich protein II (PfHRP II), *P. falciparum* parasite lactate dehydrogenase (*Pf*-pLDH), *Plasmodium vivax*–specific pLDH (*Pv*-pLDH), common human *Plasmodium* LDH (pan-pLDH), and aldolase are the important biomarkers or antigens detected in the rapid diagnostic tests available for malaria parasites. Given the extensive literature describing their utility, these biomarkers will not be described here. The chapter includes only the biomarkers of recent interest.

Hemozoin

Hemozoin is a nontoxic insoluble microcrystalline by-product formed by polymerization of free-toxic heme after digestion of hemoglobin. This malaria pigment is demonstrated inside the infected RBC and is a nonspecific feature of all *Plasmodium* spp. Ten billion parasites produce 3–4 μmol of hemozoin, which corresponds to around 0.4512 pg of hemozoin per parasite. Its presence in the blood sample indicates malaria without species differentiation. In different studies, sensitivity and specificity range from 80% to 90%. False-positive results have been observed in patients who have been successfully treated because the pigment-laden leucocytes persist in circulation for 2–3 weeks more.

Originally, detection of hemozoin was done by flow cytometer, and later by MALDI-TOF. Some commercial hematological analyzers are also equipped for hemozoin detection. In recent years, improved biosensor systems have been used for its detection, which are fast and can be used for malaria screening. One such device is Gazelle, developed by Kumar and coworkers. This system detects the LED-emitted light in the sample in a magnetic field. The quantity of light is proportional to the hemozoin concentration in the sample. According to authors, it can detect as few as 50 *P. falciparum* parasites/μL and 35 *P. vivax* parasites/μL with 95% and 100% accuracy, respectively. It delivers the result within 1 minute and costs about US$1 per test. The instrument is bulky, however, and not suitable for field conditions. Several biosensor-based methods have been evaluated for hemozoin detection, but they are still in experimental stages.

Glutamate Dehydrogenase

P. falciparum produces three glutamate dehydrogenase (GDH) enzymes, of which Type 1 (*Pf*GDH1) has been used for its detection. These enzymes are absent in RBC but are present throughout the erythrocytic cycle of the parasite. Western blot technique and immunochromatographic assay have been evaluated using monoclonal antibodies. The sensitivity and specificity of the tests have been reported to be almost 90%. An ELISA-based test has also been described using polyclonal antibodies raised against recombinant parasite *Pf*GDH.

Other Biomarkers

Parasite hypoxanthine phosphoribosyl-transferase (pHPRT) and phosphoglycerate mutase (pPGM) were found to be higher in patients with *P. falciparum* than in controls. Apical membrane antigen 1 and merozoite surface proteins 1 and 2 are some other interesting candidate molecules that have been studied.

***Leishmania* spp.**

Several parasite specific biomarkers have been studied by different groups, which are summarized in the following sections.

Kinesin-Related Proteins

Kinesin proteins in *Leishmania* spp. are part of the microtubule cytoskeleton. Of the various kinesin-related proteins, recombinant K39 (rK39) of *Leishmania infantum* has been used as a biomarker, and assays are available for detection of antibodies to this protein for diagnosis of human and canine visceral leishmaniasis. In addition, other kinesin proteins such as K26, K28 and

K9 have been studied, and their recombinant forms have been used to design immunoassays similar to rK39. Proteins like rK26 and rK28 have shown sensitivity and specificity similar to and even higher than those of rK39. Protein such asrKE16 biomarker from *Leishmania donovani* has also been used in an immunochromatography test for visceral leishmaniasis.

Heat Shock Proteins

This class of proteins are stress proteins and are highly conserved molecules. The Heat Shock Protein 70 (HSP70) from *L. donovani* and other *Leishmania* spp. have been characterized. These molecules elicit a robust antibody response, and rHSP70 and rHSP83 have been used for serodiagnosis of visceral leishmaniasis and cutaneous leishmaniasis.

Ribosomal Proteins

The large subunit contains various copies of a protein designated as P proteins. The LiP2a, LiP2b, and LiP0 proteins from *L. infantum* have been shown to produce antibodies and are being evaluated to diagnose visceral leishmaniasis in humans and canines. Some form of cross-reactivity has been observed in systemic lupus erythematosus and in Chagas disease.

Peroxiredoxins

Tryparedoxins belong to the class of oxidoreductases and are present in all developmental stages of the parasite. Two such proteins derived from *L. infantum* are LiTXN1 and LicTXNPx. These have been found to be useful for diagnosis of both human and canine infections.

Others

Cysteine proteinases, nuclear proteins, A2 protein, and rKLO8 are other novel biomarkers against which antibodies can be detected in various forms of leishmaniasis. The immunodominant protein KLO8 needs special mention. It was observed that in Sudan, where visceral leishmaniasis is endemic, the rK39 test showed poor performance, with low sensitivity. By contrast, ELISA developed with the recombinant rKLO8 protein has shown improved sensitivity. It also does not suffer from cross-reactivity in malaria infection, as has been shown with rK39 antigen.

Biomarkers for Other Parasites

Proteomic studies with various parasites are identifying novel biomarkers that can be used for immunological diagnosis. One biomarker that has received considerable attention is the heat shock protein (HSP). Parasite-specific HSPs are conserved among different species or genera, are highly immunogenic, and sera of patients recognize these proteins. Based on their molecular weights, these include HSP110, HSP90, HSP70, HSP60, HSP40, HSP20-30, and HSP10. The two HSPs that have been studied in parasites are HSP70 and HSP90.

Research in HSP90 has focused on protozoan infections including *Giardia* spp, *Leishmania* spp., *Tryapnosoma cruzi*, *P. falciparum*, and *Toxoplasma gondii*. HSP70 has been studied as a diagnostic marker for some helminthic parasites such as Schistosomas, *Echinococcus*, *Strongyloides*, *Trichinella spiralis*, *Wuchereria bancrofti*, *Brugia* spp., *Onchocerca volvulus*, *Taenia solium*, *Fasciola* spp., *Paragonimus westermani*, and *Clonorchis sinensis*. These proteins potentially play a role not only in diagnostics but also in drug development and vaccine design.

Evolving Molecular Immunodiagnostic Tests

Evolving molecular immunodiagnostic tests encompasses a variety of techniques.

PCR-ELISA

PCR-ELISA is a combination technique wherein the PCR-amplified DNA is detected by ELISA. It involves three steps: amplification, immobilization, and detection.

- *Amplification*: The gene of interest is amplified by PCR in the presence of digoxigenin/fluorescein-dUTP. The digoxigenin/fluorescein-labeled PCR product binds to a specific oligonucleotide probe labeled with biotin at 5′ end.
- *Immobilization*: The biotinylated gene product is then immobilized on a microtiter plate. The plates are coated with streptavidin, which binds to the biotin to form a strong avidin–biotin complex, which fixes the gene product on the plate.
- *Detection*: The biotinylate-amplified DNA is then detected using anti-digoxigenin/fluorescein-peroxidase conjugate and the substrate 2,2′-azino-di-3-ethylbenzthiazoline sulfonate (ABTS) or 3,3′,5,5′-tetramethylbenzidine (TMB). This produces a blue-green color that can be measured by spectrophotometer (ELISA reader).

The PCR-ELISA has got distinct advantages over conventional or real-time PCR applications. First, this system is more sensitive compared with conventional PCR in which the amplified product is analyzed by gel electrophoresis. This is because the specific hybridization and enzymatic steps increase the positive signal, which can be read with sensitive equipment. Second, semi-quantification can be done by measuring the color intensity, which is not possible when using conventional PCR. Third, this method ensures quicker detection, since the long-duration electrophoresis step is not needed. Finally, many samples can be analyzed at the same time at a lesser cost without the need for other expensive instruments such as an electrophoresis system and a gel documentation apparatus. It has been reported that while conventional PCR has limit of detection of 1–10 ng/μL of DNA, PCR-ELISA can detect as little as 0.01 ng/μL of DNA. However, PCR-ELISA is less sensitive than real-time PCR (level of detection is 0.25 pg/μL).

Application in Parasitology

PCR-ELISA has been used experimentally for detection of *Schistosoma* in stool specimens, trypanosomes of veterinary importance (*Trypanosoma vivax*, *Trypanosoma congolenese*, *Trypanosoma evansi*), as well as for the diagnosis of leishmaniasis.

Flow Cytometry

Flow cytometers, as the name suggests, are instruments that can rapidly analyze cells or particles as they flow past laser rays suspended in a fluid medium.

Principle of Flow Cytometry

In flow cytometry, the cells are first stained with a fluorochrome dye such as acridine orange, phycoerythrin, or fluorescein isothiocyanate (FITC). A suspension of stained cells is forced under pressure using gas into a plastic tubing to the flow chamber in the instrument. Using a needle, the sample is injected into a stream of normal saline called the sheath. Both the sheath and the cells come out of the flow chamber through an opening of 75 μm in size. This aligns the cells in a single file in the middle, surrounded by the saline sheath. Next, the cells pass through the laser beam. Laser is preferred because of its high intensity, stability, and monochromaticity. Argon laser is often used for FITC while Krypton is used in dual analysis systems. The suspended particles or cells, which may range in size from 0.2 to 150 μm, pass through the laser beam and scatter the light beams. The fluorescently labeled cell components are excited by the laser and emit light at a longer wavelength than the light source.

Various detectors pick up a combination of scattered and fluorescent light. These detectors are essentially photomultiplier tubes or photodiodes. One of these detectors is in line with the light beam and is used to measure forward scatter or FSC. Another detector is placed perpendicular to the stream and is used to measure side scatter (SSC). FSC can detect the cell volume whereas the SSC reflects the inner complexity of the particle such as its cytoplasmic granule content or nuclear structure.

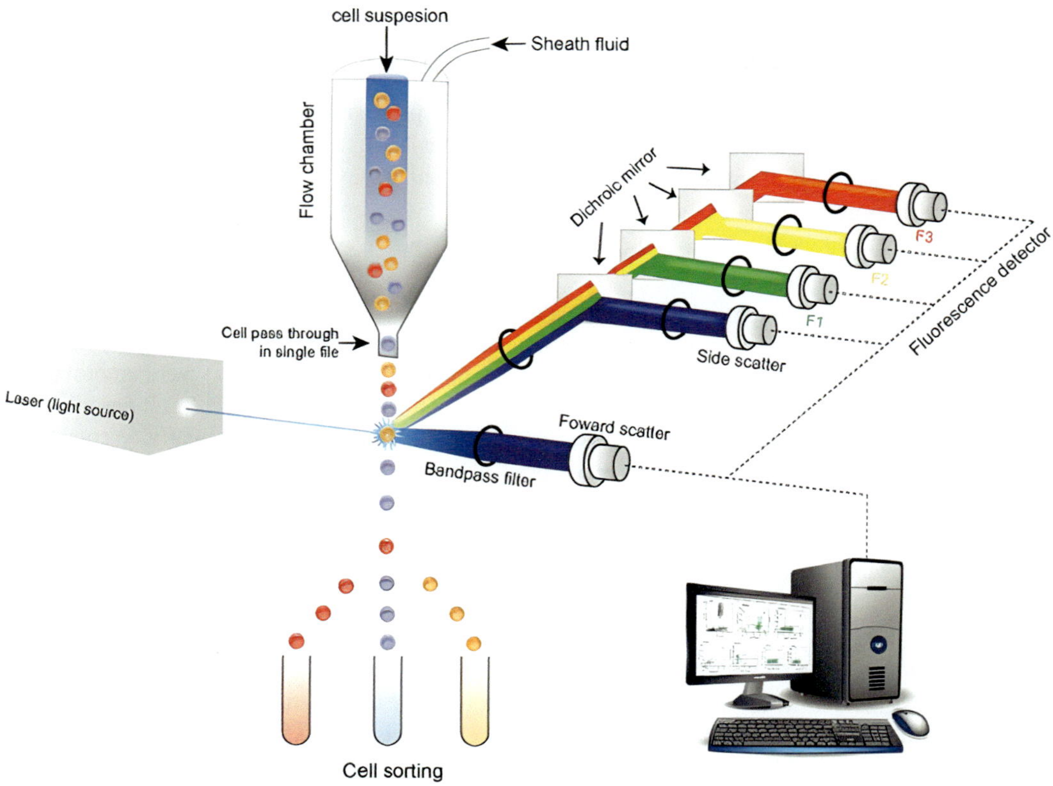

FIGURE 18.1 Principle of a flow cytometer.

(Photo credit: Creative Diagnostics, USA.)

The modern instruments contain multiple laser systems with as many as 20 parameters (FSC, SSC, and 18 fluorescent detectors), and the latest ones use five or more lasers and have 30–50 parameters (Figure 18.1). Each particle is analyzed for visible light scatter and one or multiple fluorescence parameters. The detectors transmit the information to the computer, and, using the light measurements, different information can be gathered about the physical and chemical structures of the cells. These instruments can analyze 50,000–100,000 cells per minute.

Nanomaterial-Based ELISA

In the past two decades, research in nanomaterials has resulted in the development of improved ELISA techniques that can have sensitivities many times higher than those of the conventional types. New signal readouts in ELISA using nanomaterials have been described. These can be divided into three types: substrate alternatives, enzyme alternatives, or enzyme-free amplifiers.

Nanomaterials as Substrates

In the conventional ELISA, chromogenic chemicals such as 3,3′,5,5′-tetramethylbenzidine (TMB) or o-phenylenediamine are commonly used as substrates that produce a colored product using enzyme action. These chemicals have low extinction coefficients of about 10^4/M/cm, which limits the assay's sensitivity. Plasmonic nanoparticles like that of gold (AuNP) have an extinction coefficient of 10^9/M/cm. Since the solution absorbance is directly related to extinction coefficient, the use of these nanoparticles results in higher detection sensitivity. Naked-eye detection of color change is another advantage of this assay.

This type of ELISA is referred to as plasmonic ELISA, which was first described in 2012. Except for the use of AuNP as a substrate, the basic format of ELISA remains largely unchanged. The enzyme used is catalase, which breaks down the hydrogen peroxide leading to the formation of aggregates of AuNP with red shifting due to interparticle plasmon coupling. The resultant color change is from initial sharp red to blue. Other enzymes such as glucose oxidase or alcohol dehydrogenase have also been used in place of catalase. This assay can reach a level of detection as low as 10^{-18} g/mL.

Nanomaterials as Enzymes

Nanomaterials have been explored as possible alternatives to existing enzymes such as horseradish peroxidase or alkaline phosphatase, which are amenable to pH or temperature changes. The first such report was in 2007, in which Fe_3O_4 magnetite nanoparticles (Fe_3O_4 MNPs) were employed in place of enzyme. This material has a catalytic activity similar to natural peroxidases. It has a higher affinity and a faster catalytic rate for TMB substrate and has a much higher sensitivity compared with conventional enzymes. Researchers have used it in ELISA as well as in immunochromatographic tests. These nanoenzymes lack substrate-specific binding sites and hence do not possess substrate specificity. This issue has been addressed by creating TMB- and ABTS-specific binding sites on Fe_3O_4 MNPs using a molecular imprinting technique. Newer nanoenzymes using cobalt, manganese, platinum, and copper have also been explored.

Nanomaterials as Nonenzymatic Signal Amplifiers

The methods described in the previous sections depend either on a natural enzyme or a chemical substrate. With the aim of totally replacing both natural chemicals, efforts have been made to produce an amplifier system in which no enzymatic reaction is involved but that is compatible with ELISA. Nanomaterials have been used as carrier substances that can release large amounts of signal reporters.

One such technique was first described in 2003, where detection antibody and many DNA barcodes were conjugated on the same AuNP. Combined with PCR, this Biobarcode assay reached a detection level of 3 aptomoles (aM) of the analyte. It was made possible because even trace amounts of target protein detected are expressed as detection of large numbers of DNA barcode. Apart from surface coating of nanoparticles, signal reporters can also be packaged in liposomes that are released during the detection step. The liposome-assisted plasmonic ELISA has been found to visually detect up to 6.7 aM of antibody in the sample. Similarly, copper monoxide nanoparticles can release millions of cupric ions on acid hydrolysis, which are then converted to cuprous ions by sodium ascorbate, which in turn induces alkyn- and azid-modified AuNP aggregation. This can be used as a point-of-care test.

The following sections outline the procedural information about the design of PCR-ELISA and a flow cytometry–based diagnostic test for detection of anti-*Leishmania* IgG.

Design and Optimization of PCR-ELISA Protocol for *Schistosoma mansoni*

The procedure outlined by Gomes et al. can be used to detect *Schistosoma mansoni* in stool specimens with improved sensitivity.

Reagents and Equipment Required

Reagents: bovine serum albumin; magnesium chloride; formamide; PBS-T; anti-fluorescein-horseradish peroxidase; TMB; PCR master mix buffer (commercial).

Equipment: thermal cycler; streptavidin-coated microtiter plates (commercial); ELISA reader.

Others: DNA extraction kit (commercial); biotin 5′–labeled forward primer and unlabeled reverse primer; fluorescein 5′–labeled probe; human beta actin gene; deoxynucleotide triphosphates (dNTP).

Designing the PCR-ELISA

Specimen Collection

Stool samples from suspected patients are collected and stored at −70°C until DNA extraction.

Test Procedure

DNA extraction

- DNA is extracted from 500 mg of stool using commercial DNA extraction kit (QIAamp DNA Stool Mini Kit, Quiagen, Germany).
- The protocols for DNA isolation from large amounts of stool and that for pathogen detection in stool are followed according to the kit instructions.
- Heating is done at 95°C for 20 minutes to rupture the eggs of *S. mansoni.*

DNA amplification

- Primers and probe: biotin 5′–labeled forward primer (5′-GATCTGAATCCGAC-CAACCG-3′), an unlabeled reverse primer (5′-ATATTAACGCCCACGCTCTC-3′), and the fluorescein 5′–labeled probe (5′-TGGTTTCGGAGATACAACGA-3′).
- The primers are capable of amplifying a 121 bp tandem repeat sequence present in *S. mansoni* DNA.
- Human beta actin gene is used as housekeeping gene and amplified using biotin-labeled primers and fluorescein-labeled probe.
- Fivefold dilution of the sample DNA is done and 2 μL is used as template.
- The master mix is made to a total volume of 20 μL as follows:
  Sample template DNA: 2 μL
  GeneAmp 10X PCR Gold Buffer (150 mM Tris-HCl, pH 8.0, 500 mM KCl): 2 μL
  Amplitaq Gold (Applied Biosystems): 2.0 U
  Bovine serum albumin: 0.1 mg/mL
  Each primer: 0.5 μM
  Magnesium chloride: 1.5 mM
  Each deoxynucleotide triphosphate (dNTP): 200 μM
- Before starting the PCR cycles, the mixture is heated to 95°C for 12 minutes to activate the HotStart Taq polymerase.
- The PCR cycles used are:
  95°C for 1 minute, 63°C for 1 minute, and 72°C for 30 seconds: 15 cycles
  80°C for 1 minute, 63°C for 1 minute, and 72°C for 30 seconds: 12 cycles
  80°C for 1 minute, 65°C for 1 minute, and 72°C for 30 seconds: 7 cycles
  Final elongation: 72°C for 7 minutes.
- Positive control containing DNA from a known positive *S. mansoni* sample and negative control containing all ingredients except the DNA are amplified in parallel.
- To calculate the cutoff value, samples from 20 healthy individuals are tested.

ELISA

Streptavidin-coated microtiter plates with other necessary reagents are available commercially (PCR Plate Detection Kit, Sigma-Aldrich), and the manufacturer's instructions are followed.

- 5 μL of amplified DNA is mixed with 95 μL of dilution buffer and dispensed in streptavidin-coated wells, incubated at 37°C for 30 minutes.
- Denaturation solution is added to a final volume of 100 μL and incubated at room temperature for 10 minutes followed by washing four times with PBS-Tween 20 (PBS-T).
- Plate hybridization buffer (supplied with the kit) containing 0.2 pM of fluorescein 5′–labeled probe and 20% of formamide is added to each well of the plate at a volume of 200 μL and incubated at 37°C for 45 minutes, followed by washing with PBS-T five times.

- 150 μL of anti-fluorescein-horseradish peroxidase that has been diluted 500-fold is added to each well, and further incubated at 37°C for 45 min, followed by five washes with PBS-T.
- 100 μL of TMB substrate solution is then added and the plate is kept in the dark for 5 minutes, followed by addition of 50 μl of stop solution provided in the kit.
- Absorbance is read in an ELISA reader at 450nm.

Results and interpretation: The cutoff value is calculated as the mean absorbance value of the negative controls (healthy individuals) plus three standard deviations, derived from a single PCR-ELISA test.

The sensitivity and specificity of the test are calculated based on the results of 2-slide and 12-slide Kato-Katz technique. PCR-ELISA showed a sensitivity of >96% and a specificity of >80%.

Note: Instead of using the kit as mentioned above (PCR Plate Detection Kit, Sigma-Aldrich), another kit is available from the same manufacturer that uses the digoxigenin-labeled products (PCR-ELISA [DIG Detection], Sigma-Aldrich).

Quality assurance: PCR-positive control containing DNA from a known positive *S. mansoni* sample and a negative control containing all ingredients except the DNA to be amplified in parallel with the samples.

ELISA: Known negative controls from healthy individuals to be used in the test run. The positive control amplified in PCR to be used as ELISA positive control.

Troubleshooting: A troubleshooting guide is provided with the later-mentioned kit (DIG Detection), the salient points of which are summarized in Table 18.1.

Design and Optimization of Flow Cytometry Protocol for Detection of Anti-*Leishmania* IgG

The procedure outlined here can be used to detect anti-*Leishmania* IgG as described by da Silva et al.

Reagents and Equipment Required

Reagents: PBS; fetal bovine serum (FBS); paraformaldehyde solution.

TABLE 18.1
Troubleshooting Guide for PCR-ELISA

Issue	Possible Reasons	Solution
Unexpected color development	Incubation time and/or temperature inadequate	Ensure correct time and temperature. All reagents brought to room temperature before use
	Poor water quality	Use double-distilled water free from microbial contamination
	Substrate contaminated	Pipetting directly from substrate bottle not permitted
Suspicious readings	Incorrect filters used	Filters checked for correct wavelengths
Poor precision	Sample is turbid or contains high amounts of lipids or particulate matters	Sample to be centrifuged to pellet the particles Samples to be mixed well before pipetting
	Sample or standards carry over	Pipette tips to be changed at each step
	Unequal volumes added to wells	Calibrate the pipettes
	Inadequate aspiration of fluid from wells	Fluids should be aspirated completely from wells
	Incomplete washing	Check the working of the automatic washer
	Unequal mixing of samples during incubation.	Use plate shaker

Equipment and labware: centrifuge; microtiter plate; incubator; water bath; FACS instrument (Becton Dickinson).

Others: *Leishmania infantum* promastigotes; anti-human IgG conjugated to fluorescein isothiocyanate (commercial); Cell Quest Pro software.

Designing the Flow Cytometry Protocol

Specimen Collection

Serum samples from suspected patients are used. For detection of IgG, blood is collected from the patients and centrifuged for 10 minutes at 2,800 rpm. The serum sample is then inactivated at 56°C for 30 minutes and re-centrifuged at 4°C at 2,800 rpm for 5 minutes. The supernatant is then stored at −20°C until further use.

Test Procedure

Antigen Preparation

Promastigotes of *Leishmania infantum* are used as the source of antigen.

- *L. infantum* strain MHOM/BR/70/BH46 is cultured and the parasites are harvested.
- The promastigotes are washed thrice with ice cold PBS + 10% FBS.
- The washed parasites are then incubated in 1% paraformaldehyde solution overnight.
- After final wash, it is resuspended in PBS+FBS, parasite count was made and used for the test.

Antigen–Antibody Reaction

- 2.5×10^5 parasites are dispensed in each well of a microtiter plate and different dilutions of the patient sera were added to it.
- The authors found that a serum dilution of 1:2,048 best defined the optimal positive fluorescent parasites (PPFP) in flow cytometry.
- The mixture is incubated at 37°C for 30 minutes, followed by two washes with PBS +FBS.

Conjugate Addition

- Anti-human IgG conjugated to fluorescein isothiocyanate (available commercially) at a dilution of 1:200 in PBS+FBS is added to the parasites and incubated at 37ºC for 30 minutes in the dark.
- The FITC-labeled parasites are then washed and fixed with 200 μL of 1% paraformaldehyde in the dark for 30 minutes at 4°C before flow cytometry analysis.

Flow Cytometer Analysis

- The analysis is done on FACS Calibur, Becton Dickinson, using the Cell Quest Pro software with 20,000 events per sample.
- Promastigotes are recognized based on their FSC and SSC properties of light. The parasites assume a characteristic distribution with these parameters.
- A histogram analyzes the relative fluorescence intensity of each event. The percentage of PPFP for each sample is determined using a delimitation set on the FITC-conjugated internal control histogram, which is applied to all data analysis.
- The PPFP cutoff point and the optimum dilution are then selected to get the IgG reactivity.

Results and interpretation: The authors observed that the PPFP value of 1.26 was the most appropriated cutoff to distinguish negative from positive results.

Assay performance: It was found in the study that flow cytometry analysis had a sensitivity of 89%, specificity of 83%, a 94% negative predictive value, and an accuracy of 85%.

Quality assurance: Ideally, the QC material should be similar to the specimen matrix and must have a long shelf life. The final aim is to guarantee the optimum performance of all the reagents and proper functioning of the instrument. Normally, the commercially available and approved flow cytometric methods will include QC material, but even these QC materials may be imperfect. A flow cytometer is a complex instrumentation system, and quality assurance in flow cytometry encompasses all of the following:

a. Instrument quality control
b. Reagents and antibody quality control
c. Sample and process quality control.

For details of quality control in flow cytometry, readers can consult the articles by Dorn-Beinke and Sack and by der Strate (see Further Readings section).

Commercially Available Kits and System

The diagnostic methods mentioned in this chapter are primarily experimental attempts without translation to commercial availability. One exception is the hemozoin detection system Gazelle™ marketed by Hemex Health, USA. However, the PCR-ELISA and the flow cytometry methods can be adapted for routine laboratory use, since most of the specialized reagents and the purified antigens are available commercially and need not be prepared separately.

CONCLUSION

The present century is witnessing tremendous advancements in science and technology, which are reflected in improved diagnostic techniques for infectious diseases. Many of these novel diagnostic methodologies are still in the research stage, but some of them are already making their presence felt. The primary aim of any of these improved diagnostics is to enhance the sensitivity, along with making them user friendly, which can be easily adapted as point-of-care tests. It is to be expected that nanoparticle-based technology will be more widespread in the near future, with ELISA formats gradually changing to adapt to this technology.

FURTHER READINGS

Baptista V, Peng WK, Minas G, Veiga MI, Catarino SO. Review of microdevices for hemozoin-based malaria detection. *Biosensors (Basel).* 2022;12(2):110.

Carrillo E, Moreno J. Editorial: Biomarkers in leishmaniasis. *Front Cell Infect Microbiol.* 2019;9:388.

da Silva ED, de Oliveira BC, Pereira AMdS, Guedes DL, de Melo Neto OP, Costa CHN, et al. A flow cytometry-based serological assay to detect visceral leishmaniasis in HIV-infected patients. *Front Med.* 2021;8:55328.

Dorn-Beineke A, Sack U. Quality control and validation in flow cytometry. *J Lab Med/Laboratoriums Medizin.* 2016:1. https://doi.org/10.1515/labmed-2016-0016

Gao L, Yang Q, Wu P, Li F. Recent advances in nanomaterial-enhancedenzyme-linked immunosorbent assays. *Analyst.* 2020;145:4069.

Gomes LI, Marques LHdS, Enk MJ, de Oliveira MC, Coelho PMZ, Rabello A. Development and evaluation of a sensitive PCR-ELISA system for detection of schistosomainfection in feces. *PLoS Negl Trop Dis.* 2010;4(4):e664.

Jain P, Chakma B, Patra S, Goswami P. Potential biomarkers and their applications for rapid and reliable detection of malaria. *Biomed Res Int.* 2014;2014:852645.

Kobets T, Badalová J, Grekov I, Havelková H, Svobodová M, Lipoldová M. Leishmania parasite detection and quantification using PCR-ELISA. *Nat Protoc.* 2010;5(6):1074–1080.

Kumar R, Verma AK, Shrivas S, Thota P, Singh MP, Rajasubramaniam S, Das A, Bharti PK. First successful field evaluation of new, one-minute haemozoin-based malaria diagnostic device. *E Clinical Medicine.* 2020;22:100347.

McKinnon KM. Flow cytometry: an overview. *Curr Protoc Immunol.* 2018;120:5.1.1–5.1.11.

Sue MJ, Yeap SK, Omar AR, Tan SW. Application of PCR-ELISA in molecular diagnosis. *Biomed Res Int.* 2014;2014:653014.

van der Strate B, Longdin R, Geerlings M, Bachmayer N, Cavallin M, Litwin V, et al. Best practices in performing flow cytometry in a regulated environment: Feedback from experience within the European bioanalysis forum. *Bioanalysis.* 2017;9:1253–1264.

19 Immunoelectron Microscopy

INTRODUCTION

Electron microscopy has been in use for many decades since its invention by Ernst Ruska and Max Knoll in the 1930s. It has found its use in diverse fields not only in biological sciences and medicine but also in electronics and material sciences.

In parasitic and other infectious diseases, immunoelectron microscopy has been instrumental in elucidating stage-specific antigen expression, organelle function, and host–pathogen interactions, particularly in complex intracellular parasites such as malaria.

The outstanding features of electron microscopy are its unmatched magnification and its exquisite resolution powers. The magnifying power is almost 10^6x compared with 1,500x of an optical microscope, while the resolution power is about 250 times higher (0.001 µm compared with 0.3 µm). These two properties have made electron microscopy an invaluable tool to study the complex ultrastructural details in biological specimen. It remains the method of choice for analysis of nearly every subcellular component and organelles.

Two basic types of electron microscopes used are scanning electron microscope (SEM) and transmission electron microscope (TEM). The SEM is ideal for describing the surface appearance of cells, cell–cell interaction, and topography of biofilms. The TEM has a higher resolution power and is suitable for study of cellular organelles and macromolecules.

Immunoelectron microscopy (IEM) is an offshoot of TEM whereby antigens present in a cell are recognized by using specific antibodies, which are then visualized by electron-opaque gold particles used as a probe. It provides the information on subcellular localization of an antigenic protein in a cell. The IEM is used for identifying viruses in a clinical specimen, but its role in diagnostic parasitology is limited. It can be used to characterize a hitherto unknown parasite and its taxonomic position. It remains a most useful tool for study of host–parasite interactions, identification of specific immune targets, and pinpointing the localization, transport, and expression of subcellular parasite antigens. Hence, IEM remains largely a research tool in parasitology without much contribution to routine diagnostics.

WORKING PRINCIPLES OF ELECTRON MICROSCOPY

Principle of Scanning Electron Microscopy

Scanning electron microscopy is designed to examine the surface architecture and three-dimensional topography of biological and nonbiological specimens. In SEM, a finely focused beam of high-energy electrons is generated from an electron gun and scanned in a raster pattern over the surface of the specimen. Interaction of the incident electron beam with the specimen surface leads to the emission of secondary electrons, backscattered electrons, and other signals, the intensity of which varies depending on surface contours and composition. These emitted electrons are collected by detectors and converted into electrical signals that are processed to form a detailed image of the specimen surface.

Biological specimens intended for SEM are usually fixed, dehydrated, and coated with a thin conductive layer of metal such as gold or gold–palladium to prevent charging and improve signal generation. Since the electron beam does not pass through the specimen, SEM does not provide information about internal ultrastructure. Instead, it excels at revealing surface morphology, spatial relationships between cells, cell–cell interactions, and surface specializations of parasites, including attachment organs and tegumental features.

DOI: 10.1201/9781003644224-24

Principle of Transmission Electron Microscopy

Transmission electron microscopy is based on the transmission of an electron beam through an ultrathin specimen to visualize internal ultrastructural details. The electron beam, with a wavelength of approximately 0.005 nm, is generated by an electrically heated tungsten filament under high voltage and is focused using electromagnetic lenses (Figure 19.1). The resolving power of TEM is inversely related to the wavelength of the electrons, allowing visualization of cellular structures on the nanometer scale.

Biological specimens are chemically fixed, typically using aldehydes and osmium tetroxide, dehydrated, and embedded in resin. Ultrathin sections, usually 60–90 nm thick, are cut using an ultramicrotome and mounted on copper grids. These sections are then contrasted with electron-dense heavy metal salts such as uranyl acetate and lead citrate. Differential scattering and absorption of electrons by various cellular components produce contrast within the specimen. The transmitted electrons are focused by electromagnetic lenses to form a magnified image, which is recorded by a digital camera and displayed on a monitor. TEM thus provides detailed information on cellular organelles, membranes, and macromolecular complexes.

Principle of Immunoelectron Microscopy

Immunoelectron microscopy is a specialized extension of transmission electron microscopy that combines ultrastructural imaging with immunological specificity to localize antigens at the subcellular level. The fundamental principle of IEM relies on the specific binding between an antigen and its corresponding antibody, with the immune complex visualized using electron-opaque markers.

In the IEM, antigen localization may be performed either before embedding of the specimen (pre-embedding method) or after embedding and sectioning (post-embedding method). In both approaches, a primary antibody specific to the antigen of interest is first applied. This is followed by a secondary antibody conjugated to colloidal gold particles, usually ranging from 5 to 20 nm in diameter. These gold particles appear as distinct, electron-dense dots under TEM, allowing precise localization of antigens within cellular compartments.

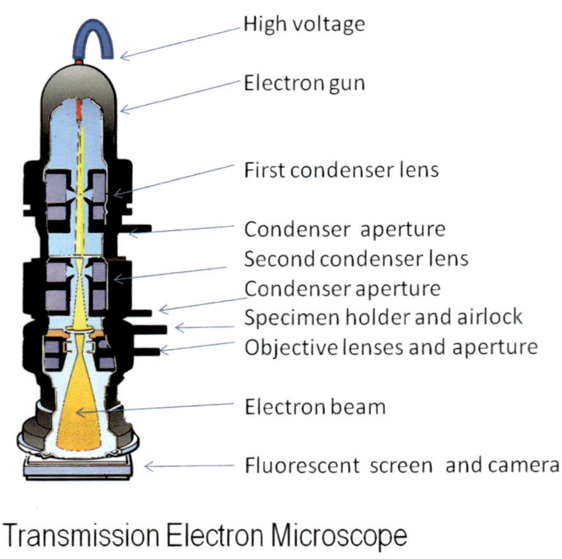

FIGURE 19.1 A cross-sectional, simplified diagram of a transmission electron microscope.

(Picture credit: Wikipedia, Dr. Graham Beards.)

To preserve antigenicity, conventional fixation and processing methods used in routine TEM are modified in IEM. The use of strong fixatives such as osmium tetroxide and glutaraldehyde is minimized or avoided, as they can mask antigenic sites. Hydrophilic embedding resins, particularly methacrylate resins, are preferred over hydrophobic epoxy resins, as they better maintain antigen–antibody reactivity. Through this approach, IEM enables correlation of structural details with molecular identity, making it a powerful research tool for studying host–pathogen interactions, parasite biology, and the intracellular distribution of specific proteins.

IMMUNOELECTRON MICROSCOPY IN INFECTIOUS DISEASES

Immunoelectron microscopy has played an important role in advancing the understanding of infectious diseases by enabling the precise ultrastructural localization of microbial antigens within host cells and tissues. IEM combines the high resolution of transmission electron microscopy with the specificity of antigen–antibody interactions. This results in direct visualization of pathogens and their antigenic components at the subcellular level. This approach has been particularly valuable in situations where conventional light microscopy or routine electron microscopy cannot reliably distinguish between morphologically similar organisms or delineate specific molecular determinants of pathogenicity.

In viral infections, IEM has been extensively used to confirm the presence of viral particles in clinical specimens, to identify intracellular sites of viral assembly, and to differentiate viruses based on antigenic composition rather than morphology alone. Immunogold labeling has enabled the localization of viral structural and nonstructural proteins within nuclei, cytoplasm, or membranous compartments, thereby contributing to a better understanding of viral replication strategies and host cell manipulation. IEM has also been applied to characterize emerging and previously unrecognized viruses during outbreaks, particularly when culture systems or molecular assays were not readily available.

In bacterial infections, IEM has been used to study the distribution of surface antigens, toxins, and virulence-associated proteins, as well as their interaction with host cell membranes and intracellular compartments. It has contributed to insights into mechanisms of bacterial adherence, invasion, intracellular survival, and immune evasion. Similarly, in fungal and parasitic infections, IEM has facilitated the localization of stage-specific and organelle-associated antigens, aiding in elucidation of host–pathogen interactions and identification of potential diagnostic or vaccine targets.

Table 19.1 summarizes the comparative roles and applications of scanning electron microscopy, transmission electron microscopy, and immunoelectron microscopy in the study and investigation of infectious diseases.

Application of IEM in Parasitology

IEM has been used for diverse applications in parasitology to have a better understanding of the surface and internal protein antigens and to find out the structure–function relationship of the antigens present in the organelles. It also potentially has a role in identification of newly discovered parasites based on the immunodominant proteins.

The bulk of the work has been carried out with malaria parasites where all the conventional approaches of electron microscopy, namely SEM, TEM, and IEM, have been exploited. This has helped generate a large amount of data about many of the parasite proteins and their functions, which was not possible with light microscopy. In the 1970s, the first reports of the binding of the antibodies to the malaria parasite surface antigens started appearing as visualized by electron microscopy. Subsequent studies used metallic labels for antibodies like ferritin and later gold particles for accurate localization of these proteins.

TABLE 19.1
Comparative Applications of SEM, TEM, and IEM in Infectious Diseases

Aspect	Scanning Electron Microscopy (SEM)	Transmission Electron Microscopy (TEM)	Immunoelectron Microscopy (IEM)
Primary focus	Surface morphology and topography	Internal ultrastructure	Ultrastructural localization of specific antigens
Type of information obtained	Three-dimensional surface details	Two-dimensional internal structural details	Structural details correlated with molecular identity
Visualization of pathogens	External morphology of bacteria, fungi, parasites, and biofilms	Internal organization of viruses, bacteria, parasites, and infected host cells	Precise localization of microbial or host antigens within pathogens or host cells
Application in viral infections	Visualization of virion surface features and budding patterns	Identification of viral particles, replication complexes, and assembly sites	Localization of viral proteins during entry, replication, and assembly
Application in bacterial infections	Study of bacterial surface structures, pili, flagella, and biofilms	Visualization of intracellular bacteria and bacterial ultrastructure	Localization of surface antigens, toxins, and virulence factors
Application in parasitic infections	Surface architecture of protozoa and helminths, attachment organs	Internal organelles and developmental stages of parasites	Localization of stage-specific and organelle-associated parasite antigens
Host–pathogen interaction studies	Cell–cell interactions and surface adherence	Intracellular interactions and host cell damage	Identification of immune targets and antigen distribution at host–pathogen interface
Diagnostic utility	Limited, mainly descriptive	Useful for confirmation of viral and parasitic infections	Very limited; mainly confirmatory or research oriented
Research applications	Morphological characterization	Ultrastructural and developmental studies	Structure–function correlation and antigen mapping
Technical complexity	Moderate	High	Very high
Requirement of antibodies	Not required	Not required	Essential (primary and labeled secondary antibodies)
Overall role in infectious diseases	Descriptive and structural	Structural and confirmatory	Molecularly specific, primarily research based

IEM has been used to detect surface-exposed antigens of extracellular stages of such parasites as merozoites, sporozoites, gametes, and ookinetes and to study altered host cell antigens expressed on the surface of infected host cells. In addition, internal antigens (e.g., in rhoptries of merozoites) have been studied by TEM and IEM. IEM has been used to study the ultrastructural details of important protozoan parasites such as *Cryptosporidium* spp., *Toxoplasma gondii*, and *Giardia* spp. Similar studies have been done for immunolocalization of parasite antigens in *Onchocerca volvulus* and to characterize the cross-reactive antigens of *Trichinella spiralis* and *Trichinella pseudospiralis* (Table 19.2).

Design and Optimization of the Protocol for Immunoelectron Microscopy for Parasites

At present, immunoelectron microscopy has no direct role in routine laboratory diagnosis of malaria. However, it has been extensively used in malaria as a research tool to elucidate the ultrastructural localization of parasite antigens and to define their functional relevance.

TABLE 19.2
Applications of Immunoelectron Microscopy in Parasitic Diseases

Parasites	Developmental Stage Studied	Antigen(s) Localized by IEM	Major Ultrastructural Site(s)	Key Application	Study Focus
Plasmodium falciparum (malaria)	Merozoite, sporozoite, trophozoite	Merozoite surface proteins, rhoptry and microneme proteins	Parasite surface, rhoptries, micronemes, dense granules	Elucidation of invasion mechanisms and stage-specific antigen expression	Host–parasite interaction, vaccine target identification
Plasmodium vivax (malaria)	Intraerythrocytic stages	Surface and exported antigens	Parasite membrane, infected erythrocyte membrane	Demonstration of antigen export to host red cell	Immune evasion, antigen trafficking
Toxoplasma gondii (toxoplasmosis)	Tachyzoite, bradyzoite	Surface and secretory antigens	Micronemes, rhoptries, dense granules	Correlation of organellar proteins with invasion and intracellular survival	Pathogenesis and virulence mechanisms
Cryptosporidium spp. (cryptosporidiosis)	Sporozoite, intracellular stages	Surface and attachment antigens	Parasite–host interface	Identification of attachment-related antigens	Host cell adherence and invasion
Giardia duodenalis (giardiasis)	Trophozoite	Variant surface proteins	Cell surface, ventral disc	Visualization of antigenic variation	Immune evasion mechanisms
Leishmania donovani	Amastigote, promastigote	Surface and intracellular antigens	Parasite membrane, phagolysosome	Localization of antigens involved in macrophage survival	Intracellular persistence
Onchocerca volvulus (onchocerciasis)	Adult worm, microfilaria	Structural and cross-reactive antigens	Cuticle, hypodermis	Characterization of immunogenic cuticular antigens	Immunopathology
Trichinella spiralis (trichinellosis)	Larval stage	Muscle-associated antigens	Nurse cell–larva complex	Identification of shared and species-specific antigens	Comparative immunology
Trichinella pseudospiralis	Larval stage	Cross-reactive antigens	Larval surface and internal structures	Differentiation from related species	Taxonomy and antigenic characterization
Schistosoma mansoni (schistosomiasis)	Schistosomula, adult worm	Tegumental antigens	Tegument and subtegument	Mapping of surface antigens exposed to host immunity	Vaccine and diagnostic antigen discovery

Reagents and Equipment Required

Reagents: primary antibodies directed against specific parasite antigens; secondary antibodies conjugated to colloidal gold particles (commonly 5–20 nm); buffered saline solutions; blocking agents such as bovine serum albumin or skim milk; mild fixatives that preserve antigenicity. Hydrophilic embedding resins, particularly methacrylate resins, are preferred to maintain antigen–antibody reactivity.

Equipment and labware: transmission electron microscope; ultramicrotome for sectioning; copper grids; incubators for antibody reactions; digital imaging systems. Access to a dedicated electron microscopy facility with trained technical personnel is essential.

Other requirements include well-characterized parasite cultures or infected erythrocyte preparations, validated antibody panels, and appropriate positive and negative controls to ensure specificity of immunolabeling.

Design and Optimization of the Test

Antigen Targets and Antibody Reagents

Immunoelectron microscopy studies in malaria have focused on antigens expressed at different developmental stages, including sporozoites, merozoites, and intraerythrocytic stages. Antibodies have been raised against merozoite surface proteins, rhoptry-associated proteins, micronemal antigens, and dense granule components. These antigens were selected based on their putative roles in erythrocyte invasion, intracellular survival, and immune recognition. Both polyclonal and monoclonal antibodies have been employed, with monoclonal antibodies providing higher specificity for individual proteins and enabling discrimination between closely related subcellular compartments.

Test Procedure

- Infected erythrocytes are fixed using low concentrations of aldehydes to preserve both ultrastructure and antigenicity.
- Specimens are embedded in suitable resins and ultrathin sections are prepared.
- Sections are incubated with stage-specific primary antibodies directed against the antigen of interest.
- After thorough washing, gold-labeled secondary antibodies are applied.
- Different sizes of colloidal gold particles may be used in parallel to localize multiple antigens within the same section.
- Antibody concentration, incubation time, and blocking conditions are optimized to minimize background labeling and enhance specificity.

Results and interpretation: Positive immunoelectron microscopy findings are indicated by the presence of discrete, electron-dense gold particles localized to specific parasite structures or membranes. The spatial distribution of gold particles allows direct correlation between antigen position and biological function.

Absence of labeling in negative controls, such as sections incubated without primary antibody or with irrelevant antibodies, confirms specificity. These observations have been central to defining the functional roles of malaria antigens and distinguishing surface-exposed immune targets from internally sequestered proteins.

Quality assurance: Quality assurance in malaria immunoelectron microscopy relies on the use of well-characterized antibodies, appropriate controls, and replication of findings across multiple parasite stages and preparations. Despite its strengths, the technique is labor intensive, technically demanding, and unsuitable for large-scale diagnostic application.

Immunoelectron microscopy in malaria has remained a powerful investigative tool that has contributed substantially to understanding parasite biology, antigen trafficking, and host–parasite interactions, more so than a method for routine clinical diagnosis.

Despite its considerable research value, the application of immunoelectron microscopy in routine diagnostic microbiology remains limited. The procedure is technically demanding, time consuming, and requires specialized instrumentation and expertise. Consequently, IEM is largely confined to reference laboratories and research settings, where it serves as a complementary tool to molecular and immunological methods for resolving complex questions related to pathogen structure, antigen expression, and intracellular localization in infectious diseases. As such, it is still largely restricted to specialized laboratories, limiting its role in routine diagnostics.

CONCLUSION

The electron microscope is an extremely complicated instrument and needs highly specialized operators to derive meaningful information from a specimen. The electron microscopy system is operated by dedicated personnel, and it will be foolhardy for an otherwise trained microscopist to venture into this area without extensive training. These units have established rules and operating procedures for different types of specimens and their processing. In addition, this is an expensive instrument, and only a limited number of research centers possess the necessary facilities. Therefore, it is advisable to consult the nearest electron microscopy center, clearly describing the type of microscopy required and the nature of the specimen to be analyzed. The specimen should then be submitted for analysis by trained personnel. The acquired images can be examined in consultation with an electron microscopy expert to generate meaningful and reliable data.

ACKNOWLEDGMENT

The authors gratefully acknowledge the contribution of Dr. Abhijit Poddar, PhD, of Mahatma Gandhi Medical Advanced Research Institute, Puducherry in writing this chapter.

FURTHER READINGS

Aikawa M, Atkinson CT. Immunoelectron microscopy of parasites. *Adv Parasitol.* 1990;29:151–214.

Bannister LH, Kent AP. Immunoelectron microscopic localization of antigens in malaria parasites. *Methods Mol Biol.* 1993;21:415–429.

Lakshman, MA. Review on application of conventional electronmicroscopy in disease diagnosis and research. *Indian JVetPathol.* 2022;46:1–12.

Winey M, Meehl JB, O'Toole ET, Giddings TH Jr.. Conventional transmission electron microscopy. *Mol Biol Cell.* 2014;25(3):319–323.

Zhao J, Yu X, Shentu X, et al. The application and development of electron microscopy for threedimensional reconstruction in life science: A review. *Cell Tissue Res.* 2024;396:1–18.

Appendix A

Designing an Immunodiagnostic Laboratory

INTRODUCTION

A laboratory unit involved with parasitic diseases is normally a part of the microbiology/infectious diseases laboratory. An immunodiagnostic laboratory may be a component of the common serology/immunology laboratory unit or part of high-volume facilities, or operated as a stand-alone parasitic immunodiagnostic unit. Notwithstanding the divisions, an immunodiagnostic laboratory, whether dealing with parasites only or as a shared facility where fungal/bacterial/viral serological investigations are also performed, follows a unified design without much variation. Some of the important considerations while designing such laboratory are location, functional areas, space, equipment, and necessary safety requirements.

LOCATION

The location of the laboratory should be within convenient distance from the area of collection of specimens for timely receipt of specimen and for the staff moving between departments. It should be easily accessible for the patients if the collection center is attached to the laboratory. However, to date, many hospitals have satellite collection centers on its premises or use the automated pneumatic tube systems for specimen delivery from diverse locations. The laboratory should be located away from public areas with easy staff access with due consideration for fire safety.

FUNCTIONAL AREAS

The following are the functional zones of a laboratory:

- *Central specimen reception*: This includes specimen registration, data entry, and specimen sorting as well as preliminary processing before they are sent to individual laboratories.
- *Specimen collection area*: It may be adjacent to the laboratory for outpatients or situated at a remote or satellite location. It should include a reception and waiting area for patients, toilets for patients (male and female), and cubicles for specimen collection equipped with workbenches, patient seating arrangement, and a handwashing facility.
- *Laboratories*: Depending on the service plan for the facility, immunodiagnostics may be part of the core laboratory. Separate larger space may be allotted if specialized parasitic immunology tests are performed.
- *Support areas*: These include cleaning/washing room; sterilization area; storage area for general supplies, flammable liquids, and corrosive reagents; refrigerated storage for reagents and diagnostic kits (a cold room in a large laboratory); and a separate freezer component for archived samples.
- *Staff areas*: These include staff rooms, office space, a meeting room, and a changing area with toilets, showers, and lockers.

LABORATORY SPACE AND CLEARANCE

The space determination primarily depends on the service delivery model. It may be an on-site laboratory providing a wide range of tests and services, an on-site provision limited to point-of-care (POC) testing for a few urgent tests, or networking of hospital laboratories across an area or region with individual specialization. The necessary area allotment for various components is depicted in Table A.1.

The layout of the laboratory should have the following relationship:

- Central specimen reception at the entry.
- Controlled access at entry points to staff and laboratory areas.
- Specimen flow will be from central specimen reception to the initial processing area or sorting area and then to the laboratory.
- Support areas should be close to the core/specialized laboratory.
- The staff zone with staff rooms, office, and meeting room can be accessed without passing through the core/specialized laboratory.
- Staff change areas located close to the entry for the staff to put on and remove personal protective equipment.

The width of the corridors and the laboratory work benches have been defined by the Clinical and Laboratory Standards Institute (CLSI) laboratory design guidelines (GP18-A2) as shown in Figure A.1. The aisles should be 1.5 meters wide and the workbenches should be 0.8 meters wide.

TABLE A.1
Area Allotment for Various Components

Space/Room	Small Laboratory Area in m^2 [number]	Medium Laboratory Area in m^2 [number]	Large Laboratory Area in m^2 [number]	Remarks
Reception				
Specimen reception	10 [1]	15 [1]	20 [1]	Receipt and data entry
Sorting/Initial Processing	10 [1]	15 [1]	20 [1]	
Specimen collection				
Reception	10 [1]	10 [1]	15 [1]	
Patient waiting area	10 [2]	10 [2]	20 [2]	
Collection bays	9 [1]	9 [2]	9 [4]	
Toilet for patients	4 [2]	4 [2]	4 [2]	
Toilet (Accessible)	6 [1]	6 [2]	6 [2]	Optional
Dirty utility room	8 [1]	8 [1]	12 [1]	
Storeroom	8 [1]	8 [1]	14 [1]	Consumables, sterile materials
Core laboratory	50 [1]	150 [1]	250 [1]	Biochemistry, immunology, hematology
Immunodiagnostic laboratory				
Laboratory	15 [1]	30 [1]	45 [1]	Enclosed with negative pressure

(Continued)

TABLE A.1 (Continued)

Space/Room	Small Laboratory Area in m^2 [number]	Medium Laboratory Area in m^2 [number]	Large Laboratory Area in m^2 [number]	Remarks
Anteroom	6 [1]	06 [1]	6 [1]	
Sterilization room	6 [1]	08 [1]	10 [1]	
Support area				
Emergency shower and eyewash area	1.5 [2]	1.5 [5]	1.5 [5]	
Washing room	8 [1]	15 [1]	15 [1]	
Cold room	6–10 [1]	6–10 [1]	6–10 [1]	Can be replaced with refrigerators
Deep freezer room	—	10[1]	10[1]	Optional
Storeroom	10–20 m^2 as per requirements. Separate rooms for general supplies and chemicals may be included			
Staff zone				
Meeting room	—	15[1]	25[2]	
Offices	Number of offices depends on the supervising personnel and lab managers. Single-person and two-person shared rooms, each 12m^2 may be included			For unit head, lab managers, and secretarial staff
Office workstation	Depends on number of technical staff. Each room may be about 5.5 m^2			For technical staff
Staff room	Shared	Shared	25 [2]	
Change room staff	—	—	20 [2]	Shower, toilets, lockers
Staff shower	3 [2]	3 [2]	—	

This is a minimum estimate and may be increased or modified to reflect the needs of individual units and according to the available space. Note: This is a representative model only. Individual organizations should design the laboratories based on resources, space available, and local regulatory guidelines. Created using the iHFG – International Health Facility Guidelines, available at http://healthfacilityguidelines.com.

EQUIPMENT

The range of equipment in a parasitic immunodiagnostics laboratory depends on the scope of its services. The minimum requirements include a general-purpose centrifuge, a bench top ELISA system (automatic washer and spectrophotometer/multimode reader), automatic pipettes, refrigerators to keep samples for a short time, a $-80°C$ freezer for positive sample archival, a bench top incubator, a water bath, an autoclave, and a hot air oven. Additional equipment may include an automated ELISA processor for big laboratories, a nephelometer, a luminometer, a rapid test card reader, a fluorescent microscope for immunofluorescent assays, and a light microscope for microscopic confirmation of agglutination tests.

SAFETY REQUIREMENTS

Parasite immunodiagnostic laboratories should be operated as a Biosafety Level 2 laboratory, which includes the following features:

- Doors with access control (Biometric).
- Handwashing sinks.
- Eyewash stations.
- Bench tops resistant to acid, alkali, moderate heat, chemicals, and water.
- Autoclave.

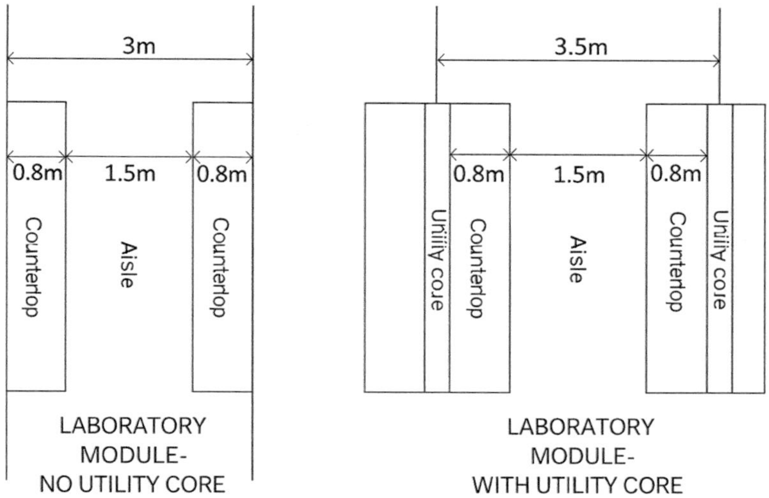

FIGURE A.1 Minimum width of aisles and workbenches as per CLSI (GP18-A2) guidelines.

- Floors should be tiled with vinyl composite or rubber or be monolithic with sheet vinyl. They should be easy to clean.
- Ideally floors and walls should be antistatic, heat resistant, antibacterial, antifungal, and chemical resistant.
- No carpeting or fabrics.
- No screens on windows, if operable. Modern laboratories do not have operable windows, since they interfere with an HVAC system.
- Negative-pressure rooms with 100% outside exhaust.
- Biosafety cabinets (may be optional).

In addition to the aforementioned, the following should be considered while designing a laboratory:

Fire Safety: Clinical laboratories fall under the low fire hazard category based on the types and quantities of flammable liquids present. The maximum amounts that can be stored in the laboratory depend on whether the laboratory is provided with sprinklers or not, the amounts being much lower in a non-sprinkler laboratory. For details of the actual amounts, the CLSI document can be consulted. Flammable liquids should not be kept near Bunsen burners, ovens, other heat sources, or in the corridors.

The walls of the laboratory in a non-sprinkler facility should have a 1-hour fire rating, which means that a fire within the laboratory should be contained in the space for 1hour.

Fire exit guidelines to help the staff exit the facility in case of a fire is a critical component. The corridors used for exit should not contain flammable or combustible items and should be minimum 6 feet (183 cm) in width. Two or more exits should be there depending on the size of the laboratory and number of personnel.

Doors of the laboratory should be of the swinging type, opening in the direction of travel, and not of the self-closing type. The main door should be at least 4 feet wide located not more than 75 feet of travel distance from any part of the laboratory.

Fire alarms and extinguishers should be available in each laboratory. Extinguishers should be situated near the exits.

Handwashing, Eye Wash, and Showers: Taps in the hand basins should be either elbow or foot operated or automatic sensor based. These handwashing stations should be provided at specimen collection area, at each subunit of the laboratory, at the automated processing area, and in the cleanup rooms. In addition, each laboratory unit must have access to at least one emergency shower and eyewash station.

The criteria for an eyewash station are as follows:

- The eyewash unit should remain turned on by itself until someone turns it off to allow the affected person to keep both eyes open while washing.
- It should turn on immediately.
- The height should be 33–45 inches from the floor.
- It should be situated not more than 100 feet, or 10 seconds of walking time, from the hazardous area.
- The supplied water should be lukewarm, not cold.
- The eyewash unit attached to the sink cannot replace fixed eyewash stations.

The criteria for emergency showers are similar to the eyewash unit except that the height should be 82–96 inches from the floor.

HEATING, VENTILATION, AND AIR CONDITIONING

The optimum temperature of a laboratory should be 20°C–22°C with a humidity of 40%–50%. The size and number of air-handling units depend on multiple factors such as the size of the laboratory, the number of persons at any one time, and heat production by the instruments, among others. The constant-air-volume type of handler is most common, which supplies a constant flow of air, in contrast to the variable-air-volume type where air volume can be regulated depending on the requirements. Whatever the type of air handler, 100% outside exhaust is essential. In a general laboratory, 6 to 10 air changes per hour are desirable. Air movement should be from a clean to a less clean area. The need for directional air flow depends on the type of laboratory unit and may not be critical for an immunodiagnostic laboratory.

CONCLUSION

Designing a clinical laboratory is a complex process that requires careful consideration of numerous criteria and strict regulatory guidelines. It is a multispecialty process involving the laboratory personnel, engineers, and architects. In addition, individual country and locality guidelines should be followed, along with the CLSI and the International Health Facility guidelines, to arrive at a proper and clear understanding.

FURTHER READINGS

Clinical and Laboratory Standards Institute. *Laboratory Design: Approved Guideline—CLSI document GP18-A2*. 2nd ed. Wayne, PA: CLSI; 2007.

Darafarin S, Movahedinia S, Darafarin H, Farzami M.Design of physical spaces in medical laboratories. *Iran J Med Microbiol.* 2021;15:480–506.

Hardy DJ. Practical aspects and considerations when planning a new clinical microbiology laboratory. *Clin Lab Med.* 2020;40:421–431.

International Health Facility Guidelines. Part B—health facility briefing & design: 145 laboratory unit, 2025. Avaialable at: https://healthfacilityguidelines.com. Accessed September 11, 2025.

Appendix B

Diagnostic Commercial Immunological Kits for Parasitic Infections

INTRODUCTION

Commercial immunological diagnostic kits for parasitic diseases are much fewer compared with other bacterial or viral infections. Even among the parasitic diseases, commercial suppliers for protozoan infections outnumber those for helminthic diseases. Most of the kit manufacturers rely heavily on the production of ELISA and rapid immunochromatographic point-of-care tests, compared with other formats, for various reasons. In this appendix, we have compiled a list of commercially available kits used for common parasitic diseases (Tables B.1, B.2, and B.3). While comprehensive, this list is not a complete one because there may be smaller manufacturing units in various countries producing similar or other kits. The readers are encouraged to contact their local jurisdictions for country- and locality-specific vendors.

TABLE B.1
Commercial Kits for Immunological Diagnosis of Parasitic Infections (Except ELISA and Rapid Tests)

Test Format	Parasite/Disease	Format	Manufacturer, Country	Website
Agglutination	Amoebiasis Schistosomiasis Toxoplasmosis Echinococcosis Distomiasis	Indirect hemagglutination	ELITech Group, France	https://www.elitechgroup.com
	Chagas Disease		Wiener Laboratorios, Argentina	https://www.wiener-lab.com.ar
			Laboratórios Lemos SRL, Argentina	https://www.lab-lemos.com
			Wama Diagnóstica, Brazil	https://www.wamadiagnostica.com.br
	Toxoplasmosis	Latex agglutination	ELITechGroup, France	https://www.elitechgroup.com
			Tulip Diagnostics, India	https://www.tulipgroup.com
			Wiener Laboratorios, Argentina	https://www.wiener-lab.com.ar
			Atlas Medical, Germany	https://atlas-medical.com
	Trypanosoma brucei gambiense	Card agglutination (CATT)	Institute of Tropical Medicine, Belgium	https://www.itg.be/en
Immunofluorescence Assay	*Giardia* antigen	Direct IF	CeLLabs, Australia	http://www.cellabs.com.au
			Meridian Biosciences, USA	https://www.meridianbioscience.com
	Cryptosporidium antigen	Direct IF	CeLLabs, Australia	http://www.cellabs.com.au
			Meridian Biosciences, USA	https://www.meridianbioscience.com
	Plasmodium antigen	Direct IF	CeLLlabs, Australia	http://www.cellabs.com.au
	Toxoplasma antibody	Indirect IF (IgG)	Hemagen Diagnostics, USA	http://www.hemagen.com
			Mast Group, Germany	https://www.mast-group.com
	Leishmania antibody	Indirect IF (IgG)	Vircell, Spain	https://www.vircell.com
		Indirect IF (IgG)	BioMérieux, France	https://www.biomerieux.com
		Indirect IF (IgG)	Euroimmun, Germany.	https://www.euroimmun.com
	Trypanosoma cruzi	Indirect IF (IgG+IgM)	Vircell, Spain	https://www.vircell.com
			Biocientífica SA, Argentina	https://biocientifica.com.ar

Immunoblot	*Trypanosoma cruzi* IgG	Western blot	LDBIO Diagnostics, France	https://ldbiodiagnostics.com/homepage
	Cysticercosis (*Taenia solium*) IgG		LDBIO Diagnostics, France	https://ldbiodiagnostics.com/homepage
			Immunetics, USA	https://immunetics.com
	Echinococcus IgG		LDBIO Diagnostics, France	https://ldbiodiagnostics.com/homepage
			Euroimmun, Germany	https://www.euroimmun.com
	Fasciola hepatica IgG		LDBIO Diagnostics, France	https://ldbiodiagnostics.com/homepage
	Toxoplasma IgG		LDBIO Diagnostics, France	https://ldbiodiagnostics.com/homepage
	Toxoplasma IgG, IgM		LDBIO Diagnostics, France	https://ldbiodiagnostics.com/homepage
	Leishmania IgG		LDBIO Diagnostics, France	https://ldbiodiagnostics.com/homepage
	Schistosomiasis IgG		LDBIO Diagnostics, France	https://ldbiodiagnostics.com/homepage
	Toxocariasis IgG		LDBIO Diagnostics, France	https://ldbiodiagnostics.com/homepage/
			Euroimmun, Germany	https://www.euroimmun.com
	Trichinellosis IgG		LDBIO Diagnostics, France	https://ldbiodiagnostics.com/homepage
Chemiluminiscence immunoassay	Toxoplasmosis IgG, IgM, IgG Avidity	Magnetic particle based	Euroimmun, Germany	https://www.euroimmun.com
	Toxoplasmosis IgG	Coated Microwell	Grupo mexlab, Mexico	https://grupomexlab.com
	Toxoplasmosis IgG, IgM	Microparticle based	Abbott Laboratories, USA, Germany	https://www.corelaboratory.abbott
	Toxoplasmosis IgG, IgM	Coated microwell	Tulip Diagnostics, India	https://www.tulipgroup.com
	Toxoplasmosis IgG, IgM, IgG Avidity	———	Diasorin, Italy	https://int.diasorin.com/en
	Plasmodium antigen	Coated microwells	Tulip Diagnostics, India	https://www.tulipgroup.com
	Trypanosoma cruzi (Chagas disease)	Microparticle based	Abbott Laboratories, USA, Germany	https://www.corelaboratory.abbott
			Roche Diagnostics	https://diagnostics.roche.com
			Diasorin, Italy	https://int.diasorin.com/en
			Vircell, Spain	https://www.vircell.com
	Giardia antigen (stool)	Magnetic particle based	Epitope Diagnostics, USA	https://www.epitopediagnostics.com

TABLE B.2
Commercial ELISA Kits for Immunological Diagnosis of Parasitic Infections

Parasite/Disease	Manufacturer, Country	Website
Antigen detection		
Entamoeba histolytica (stool)	Creative Diagnostics, USA	https://www.creative-diagnostics.com
	Tech Lab, USA	https://www.techlab.com
	CeLLabs, Australia	http://www.cellabs.com.au
	Diagnostic Automation, USA	https://diagnosticautomationinc.com
	Gold Standard Diagnostics, Germany (formerly Novatec)	https://clinical.goldstandarddiagnostics.com
Giardia intestinalis (stool)	Tech Lab, USA	https://www.techlab.com
	CeLLabs, Australia	http://www.cellabs.com.au
	apDia, Belgium	https://apdiagroup.com
	Diagnostic Automation, USA	https://diagnosticautomationinc.com
	Gold Standard Diagnostics, Germany (formerly Novatec)	https://clinical.goldstandarddiagnostics.com
Cryptosporidium spp. (stool)	Tech Lab, USA	https://www.techlab.com
	Gold Standard Diagnostics, Germany (formerly Novatec)	https://clinical.goldstandarddiagnostics.com
	R-Biopharm AG, Germany	https://clinical.r-biopharm.com
	Diagnostic Automation, USA	https://diagnosticautomationinc.com
Plasmodium spp. (blood)	Creative Diagnostics, USA	https://www.creative-diagnostics.com
	CeLLabs, Australia	http://www.cellabs.com.au
	J Mitra, India	https://jmitra.co.in
	Tulip Diagnostics, India	https://www.tulipgroup.com
Blastocystis hominis (stool)	Creative Diagnostics, USA	https://www.creative-diagnostics.com
Wuchereria bancrofti (blood)	CeLLabs, Australia	http://www.cellabs.com.au
Cysticercosis (*Taenia solium*) (blood, both human and porcine)	apDia, Belgium	https://apdiagroup.com
	Fisher Scientific, USA	https://www.fishersci.com
Dirofilariasis (heartworm)	Secure Diagnostics, India	https://www.securediagnostics.in
Animal antibody detection		
Entamoeba histolytica	Creative Diagnostics, USA	https://www.creative-diagnostics.com
	Bordier Affinity Products, Switzerland	http://www.bordier.ch
	Diagnostic Automation, USA	https://diagnosticautomationinc.com
	Gold Standard Diagnostics, Germany (formerly Novatec)	https://clinical.goldstandarddiagnostics.com
Giardia intestinalis	Equi (Ekvitestlab LLC), Ukraine	https://equitest.com.ua/en
Cryptosporidium spp.	MyBioSource, USA	https://www.mybiosource.com
	Abbexa, UK	https://www.abbexa.com
Trypanosoma cruzi	Creative Diagnostics, USA	https://www.creative-diagnostics.com
	CeLLabs, Australia	http://www.cellabs.com.au
	DiaPro, Italy	https://www.diapro.it
	Wiener Laboratorios, Argentina	https://www.wiener-lab.com.ar
	Diagnostic Automation, USA	https://diagnosticautomationinc.com
	Gold Standard Diagnostics, Germany (formerly Novatec)	https://clinical.goldstandarddiagnostics.com
	Euroimmun, Germany	https://www.euroimmun.com
Trichomonas vaginalis	Equi (Ekvitestlab LLC), Ukraine	https://equitest.com.ua/en

(*Continued*)

TABLE B.2 (Continued)

Parasite/Disease	Manufacturer, Country	Website
Leishmania spp.	Creative Diagnostics, USA	https://www.creative-diagnostics.com
	Diagnostic Automation, USA	https://diagnosticautomationinc.com
	Bordier Affinity Products, Switzerland	http://www.bordier.ch
	Euroimmun, Germany	https://www.euroimmun.com
	Secure Diagnostics, India	https://www.securediagnostics.in
Toxoplasma gondii (IgG, IgM, IgG avidity)	Creative Diagnostics, USA	https://www.creative-diagnostics.com
	Euroimmun, Germany	https://www.euroimmun.com
	Gold Standard Diagnostics, Germany (formerly Novatec)	https://clinical.goldstandarddiagnostics.com
	BiomerieuxVidas (ELFA), France	https://www.biomerieux.com
	Bio-Rad (Platelia range), USA	https://www.bio-rad.com
	Tulip Diagnostics, India	https://www.tulipgroup.com
Plasmodium spp.	Euroimmun, Germany	https://www.euroimmun.com
	Creative Diagnostics, USA	https://www.creative-diagnostics.com
	Trinity Biotech, Ireland	https://trinitybiotech.com
	NewBio, UK	https://www.new-bio.com
	DiaPro, Italy	https://www.diapro.it/
	Gold Standard Diagnostics, Germany (formerly Novatec)	https://clinical.goldstandarddiagnostics.com
	CeLLabs, Australia	http://www.cellabs.com.au
	Bio-Rad (Platelia range), USA	https://www.bio-rad.com
Babesia	Secure Diagnostics, India	https://www.securediagnostics.in
Strongyloides stercoralis	Creative Diagnostics, USA	https://www.creative-diagnostics.com
	Bordier Affinity Products, Switzerland	http://www.bordier.ch
	Euroimmun, Germany	https://www.euroimmun.com
	InBios, USA	https://inbios.com
	Diagnostic Automation, USA	https://diagnosticautomationinc.com
	Gold Standard Diagnostics, Germany (formerly Novatec)	https://clinical.goldstandarddiagnostics.com
Trichinella spp.	Creative Diagnostics, USA	https://www.creative-diagnostics.com
	Equi (Ekvitestlab LLC), Ukraine	https://equitest.com.ua/en
	Diagnostic Automation, USA	https://diagnosticautomationinc.com
	Gold Standard Diagnostics, Germany (formerly Novatec)	https://clinical.goldstandarddiagnostics.com
	Bordier Affinity Products, Switzerland	http://www.bordier.ch
	Euroimmun, Germany	https://www.euroimmun.com
	Secure Diagnostics, India	https://www.securediagnostics.in
Filariasis	Creative Diagnostics, USA	https://www.creative-diagnostics.com
	Bordier Affinity Products, Switzerland	http://www.bordier.ch
	CeLLlabs, Australia	http://www.cellabs.com.au
Toxocara canis	Creative Diagnostics, USA	https://www.creative-diagnostics.com
	Equi (Ekvitestlab LLC), Ukraine	https://equitest.com.ua/en
	Bordier Affinity Products, Switzerland	http://www.bordier.ch
	CeLLlabs, Australia	http://www.cellabs.com.au
	Gold Standard Diagnostics, Germany (formerly Novatec)	https://clinical.goldstandarddiagnostics.com
	Euroimmun, Germany	https://www.euroimmun.com

(Continued)

TABLE B.2 (Continued)

Parasite/Disease	Manufacturer, Country	Website
Ascaris spp.	Creative Diagnostics, USA	https://www.creative-diagnostics.com
	Bordier Affinity Products, Switzerland	http://www.bordier.ch
	Gold Standard Diagnostics, Germany (formerly Novatec)	https://clinical.goldstandarddiagnostics.com
	Equi (Ekvitestlab LLC), Ukraine	https://equitest.com.ua/en
Angiostrongylus cantonensis	Creative Diagnostics, USA	https://www.creative-diagnostics.com
Anisakidae	Bordier Affinity Products, Switzerland	http://www.bordier.ch
Taenia solium	Creative Diagnostics, USA	https://www.creative-diagnostics.com
Spirometra mansoni	Creative Diagnostics, USA	https://www.creative-diagnostics.com
Echinococcus granulosus	Creative Diagnostics, USA	https://www.creative-diagnostics.com
	Bordier Affinity Products, Switzerland	http://www.bordier.ch
	Equi (Ekvitestlab LLC), Ukraine	https://equitest.com.ua/en
	Euroimmun, Germany	https://www.euroimmun.com
	Gold Standard Diagnostics, Germany (formerly Novatec)	https://clinical.goldstandarddiagnostics.com
Echinococcus multilocularis	Bordier Affinity Products, Switzerland	http://www.bordier.ch
Paragonimus westermanii	Creative Diagnostics, USA	https://www.creative-diagnostics.com
Clonorchis sinensis	Creative Diagnostics, USA	https://www.creative-diagnostics.com
Opisthorchis felineus	Equi (Ekvitestlab LLC), Ukraine	https://equitest.com.ua/en
Schistosoma spp.	Creative Diagnostics, USA	https://www.creative-diagnostics.com
	Euroimmun, Germany	https://www.euroimmun.com
	Diagnostic Automation, USA	https://diagnosticautomationinc.com
	Gold Standard Diagnostics, Germany (formerly Novatec)	https://clinical.goldstandarddiagnostics.com
	Bordier Affinity Products, Switzerland	http://www.bordier.ch
Fasciola hepatica IgG/IgM	Creative Diagnostics, USA	https://www.creative-diagnostics.com
	Bordier Affinity Products, Switzerland	http://www.bordier.ch
	Diagnostic Automation, USA	https://diagnosticautomationinc.com
Fasciola gigantica	Diagnostic Automation, USA	https://diagnosticautomationinc.com

TABLE B.3
Commercial Rapid Diagnostic Test Kits for Immunological Diagnosis of Parasitic Infections

Parasite/Disease	Manufacturer, Country	Website
Malaria parasite antigen detection	Access Bio, Inc, USA	https://accessbio.net/
	CTK Biotech, USA	https://ctkbiotech.com/
	R-Biopharm, Germany	https://r-biopharm.com/
	Orchid Biomedical Systems, India	https://www.orchidbiomedical.com
	Premier Medical Corporation, India	https://www.premiermedcorp.com
	Genomix Molecular Diagnostics Pvt, India	https://genomixbiotech.com/rdt
	Zephyr Biomedicals, India	https://www.tulipgroup.com/zephyr.html
	Advy Chemical Pvt Ltd, India	https://advychemical.com/
	Pathkits, India	https://pathkits.com
	Meril Diagnostics, India	https://www.merillife.com
	Arkray Healthcare (formerly Span Diagnostics), India	https://www.arkray.co.in
	SD Biosensor healthcare, India	https://www.sdbiosensor.co.in
	Bio-Rad, USA	https://www.bio-rad.com
	Apacor, UK	https://apacor.com
	Wondfo Biotech Co. Ltd, China	https://en.wondfo.com
	InTec Products, Inc, China	https://www.intecasi.com
	Zhejiang Orient Gene Biotech, China	https://www.orientgene.com
	Abbott (Binaxnow Series), USA	https://www.globalpointofcare.abbott
	Abbott (Bioline Series), USA	https://www.globalpointofcare.abbott
	Abbott (NxTek Eliminate Malaria), USA	https://www.globalpointofcare.abbott
	RapiGEN Inc, Korea	https://www.rapigen.com
	R-Biopharm, Germany	https://clinical.r-biopharm.com
Entamoeba histolytica, Giardia, Cryptosporidium (individual/combinations) antigen detection	Meridian Bioscience, USA	https://www.meridianbioscience.com
	Thermo Fisher Scientific (Remel™), USA	https://www.thermofisher.com
	Techlab, USA	https://www.techlab.com
	R-Biopharm, Germany	https://r-biopharm.com/
	Coris Bioconcept, Belgium	https://www.corisbio.com/products/
Trypanosoma brucei gambiense antibody detection	Coris Bioconcept, Belgium	https://www.corisbio.com/products/
	Abbott (Bioline Series), USA	https://www.globalpointofcare.abbott
Trypanosoma cruzi antibody detection	Abbott (Bioline Series), USA	https://www.globalpointofcare.abbott
	InBios International, USA	https://inbios.com
	Chembio Diagnostic Systems, USA	https://chembio.com
	Wiener Laboratorios, Argentina	https://www.wiener-lab.com.ar
	Lemos SRL, Argentina	https://www.lab-lemos.com
	Cypress Diagnostics, Belgium	https://diagnostics.be
	CTK Biotech, USA	https://ctkbiotech.com/
	Bio-Manguinhos, Brazil	https://fiocruz.br
	Fujirebio, Japan	https://www.fujirebio.com
	Operon S.A, Spain	https://operondx.com
	LDBIO Diagnostics, France	https://ldbiodiagnostics.com/homepage
Toxoplasma gondii antibody detection	CTK Biotech, USA	https://ctkbiotech.com/
	Atlas Medical, Germany	https://atlas-medical.com
	Lab-Care Diagnostics, India	www.labcarediagnostics.com
	QuickLab, India	https://quicklab.in
	Tulip Diagnostics, India	https://www.tulipgroup.com
	InBios International, USA	https://inbios.com

(*Continued*)

TABLE B.3 (Continued)

Parasite/Disease	Manufacturer, Country	Website
Leishmania antibody detection	Bio-Rad, France, USA	https://www.bio-rad.com
	CTK Biotech, USA	https://ctkbiotech.com/
	Apacor, UK	https://apacor.com
	Arkray Healthcare (formerly Span Diagnostics), India	https://www.arkray.co.in
Trichomonas antigen	Sekisui Diagnostics, USA	https://sekisuidiagnostics.com
Filariasis antibody/antigen detection	Abbott (Bioline series) (IgG4 detection), USA	https://www.globalpointofcare.abbott
	Reszon Diagnostics (IgG4 detection), Malaysia	https://reszonics.com
	ICT International, South Africa	www.ictdiagnostics.co.za
	BTNX, Canada	https://www.btnx.com
	Athenese-Dx, India	https://athenesedx.com
	Pathkits, India	https://pathkits.com
	Nulife care, India	https://nulifecare.in
	SD Biosensor, South Korea	https://www.sdbiosensor.com
	Abbott (Bioline series), USA	https://www.globalpointofcare.abbott
	Drugs & Diagnostics for Tropical Diseases, USA	https://www.ddtd.org
Onchocerciasis antibody detection	Abbott (Bioline series), USA	https://www.globalpointofcare.abbott
	Drugs & Diagnostics for Tropical Diseases, USA	https://www.ddtd.org
Loa loa Antibody detection	Drugs & Diagnostics for Tropical Diseases, USA	https://www.ddtd.org
Echinococcus granulosus antibody detection	Vircell, Spain	https://www.vircell.com

Appendix C

Commercial Sources of Reagents and Chemicals

INTRODUCTION

This appendix lists commercial manufacturers of various biologicals, chemicals, reagents, and laboratory glass- and plasticware available through multinational and Indian suppliers. The availability of ready-made high-quality antigens and antibodies, along with other components such as enzyme conjugates needed to develop in-house immunological tests, eliminates the complex and time- and labor-consuming task of producing the same in the laboratory. We have tried our best to include as many as possible, but there may be some gaps, and some of the items and manufacturers may have been inadvertently missed. The readers are encouraged to inquire with the vendors in their respective countries about availability (Table C.1).

TABLE C.1
Commercial Sources of Reagents and Chemicals

Biologicals & Chemicals	Parasite	Supplier	Website
Antigens	*Cryptosporidium parvum* *Echinococcus granulosus* *Giardia lamblia* *Leishmania* *Plasmodium* *Schistosoma* *Toxoplasma gondii* *Trichomonas vaginalis* *Trypanosoma* *Dirofilaria immitis* (heartworm)	Creative-biolabs, USA	https://www.creative-biolabs.com
	Plasmodium *Echinococcus granulosus* *Leishmania* *Trypanosoma brucei* *Toxoplasma gondii* *Trichomonas vaginalis* *Acanthamoeba* *Cryptosporidium parvum* Lymphatic filariasis *Schistosoma japonicum* *Schistosoma mansoni*	Creative Diagnostics, USA	https://www.creative-diagnostics.com
	Echinococcus granulosus *Toxoplasma gondii*	Virion-Serion, Germany	https://www.serion-immunologics.com

(Continued)

TABLE C.1 (Continued)

Biologicals & Chemicals	Parasite	Supplier	Website
	Babesia canis *Babesia gibsoni* *Trypanosoma cruzi* *Dirofilaria immitis* *Leishmania infantum* *Toxoplasma gondii* *Neospora caninum*	Rekom Biotech, Spain	https://www.rekombiotech.com
	Wide range available	ThermoFisher Scientifics (Invitrogen products), USA	https://www.thermofisher.com
Antibodies Primary/ secondary labeled or unlabeled	Wide range available	ThermoFisher Scientifics (Invitrogen products), USA	https://www.thermofisher.com
	Wide range available	Vector Laboratories, USA	https://vectorlabs.com
	Wide range available	Abcepta, USA	https://www.abcepta.com
	Wide range available	ICL, USA	https://www.icllab.com
	Wide range available	Merck (Sigma Aldrich products)	https://www.sigmaaldrich.com
	Wide range available	Creative Diagnostics, USA	https://www.creative-diagnostics.com
	Wide range available	Abcam, UK	https://www.abcam.com
	Wide range available	Virion-Serion, Germany	https://www.serion-immunologics.com
	Wide range available	Cytiva, USA (Amersham products)	https://www.cytivalifesciences.com
Chemicals	ThermoFisher Scientifics and Merck (Sigma Aldrich products, Millipore products) are the leading suppliers of all chemicals related to life sciences. In addition, there are numerous manufacturers based in other countries. Indian suppliers include, among others, HiMedia, Sisco Research Laboratories (SRL), Loba-Chemie, and Central Drug House (CDH) limited.		
Glassware and Plasticware	Bio-Rad, ThermoFisher Scientifics (Nunc products, DWK Life Sciences products), and Corning are some of the leading brands of glassware, plasticware, and other laboratory consumables. Indian manufacturers include Tarsons, Borosil Limited, HiMedia, Labindia Analytical, and several others.		

Appendix D

Indian Academy of Tropical Parasitology (IATP) Guidelines for Parasitic Disease Diagnosis

INTRODUCTION

The Indian Academy of Tropical Parasitology, the brain child of the eminent medical parasitologist Prof. Subhash Chandra Parija, is a scientific society devoted exclusively to the advancement of knowledge and education in medical parasitology. Founded in 2007, the society has been active in regularly holding national and regional conferences along with the publication of its Scopus- and PubMed-indexed flagship journal, *Tropical Parasitology*. In recent years, the Academy has taken up the task of formulating diagnostic guidelines for important parasitic diseases using feedback from prominent parasitologists received at a panel discussion and symposium held at annual national conferences. The consensus opinion emerging from these meetings is declared as the Guidelines for the Laboratory Diagnosis of the particular parasitic infection and subsequently published in the *Tropical Parasitology* journal as "Declaration" named after the place of the conference. Here we have included two such declarations published in the journal in 2019 and 2024 on the laboratory diagnosis of amebiasis and paragonimiasis.

Pondicherry Declaration on the Identification and Detection of *Entamoeba histolytica*

Diagnosis of Intestinal Amoebiasis

- *Microscopy*: It is of limited utility since the cysts of *Entamoeba histolytica* are indistinguishable from other non-pathogenic members, namely *Entamoeba dispar* and *Entamoeba moshkovskii*.
- *Immunodiagnosis*: Detection of *E. histolytica* antibodies is useful for diagnosis in developed countries, but it has no use in amoebiasis endemic countries where seroprevalence is high in healthy subjects. In such cases detection of faecal antigen is a useful test and it is available in ELISA as well as rapid immunochromatographic test formats.
- *Molecular Diagnosis*: PCR detection of parasitic 18SrDNA with species-specific probe in faecal samples has high sensitivity and specificity. It can be done by conventional, nested, or real time PCR methods and can distinguish *E. histolytica* from other species.

Diagnosis of Extra-Intestinal Amoebiasis

- *Microscopy*: It has very limited role in diagnosis of conditions like amoebic liver abscess where the number of trophozoites in aspirated pus is extremely low.
- *Immunodiagnosis*: Similar to intestinal amebiasis, antibody detection in patient's serum is not useful in endemic countries. However, a negative test can rule out intestinal or extra-intestinal amebiasis. Antigen detection in blood or aspirated pus and other body fluids by ELISA can be attempted. But the commercial antigen detection tests suffer from variable sensitivities and are not recommended.

- *Molecular diagnosis*: Definitive identification of pathogenic *E. histolytica* is essential for proper management of patients and epidemiological studies during an outbreak situation. PCR assays targeting the 18SrDNA is highly specific and sensitive and are recommended in diagnosis of any form of extra-intestinal amoebiasis.

Guwahati Declaration on the Guidelines for the Diagnosis of Endemic Haemoptysis

Pulmonary paragonimiasis caused by *Paragonimus westermani*, *P. heterotremus*, *P. kallicoti*, *P. skrjabini* and other species closely mimics tuberculosis including physical signs and symptoms and the radiological feature. Sometimes it is referred to endemic haemoptysis because of the characteristic symptom in endemic regions of the world. It is imperative to distinguish it from pulmonary tuberculosis since a misdiagnosis can lead to months of anti-tuberculosis therapy without any benefit.

- *Microscopy*: Detection of the characteristic eggs by microscopy in stool or early morning sputum is considered the gold standard for diagnosis. Direct wet mount or saline mount is enough without any other staining methods. Stool examination is useful in children who tend to swallow the sputum. Multiple samples should be examined (up to 7 for sputum and 3 for stool) before giving a negative report. The main drawback is low sensitivity due to scanty number of eggs and irregular shedding.
- *Immunodiagnosis*: Antibody detection is invaluable in both pulmonary as well as extra-pulmonary paragoniasis. In pulmonary paragoniasis, antibody becomes positive 6–8 weeks before eggs can be detected and thus particularly useful in early diagnosis. ELISA and Western Blot tests can be used as screening or for confirmation and have high sensitivity and specificity. Excretory-secretory (ES) antigen-based tests are more specific compared to older crude antigen based serological tests. The 27 kDa ES antigen protein of *P. westermani* and 31.5 kDa of ES protein of *Paragonimus heterotremus* are in common use in various commercial tests. A rapid test kit (Dot-immunogoldfiltration assay) has also been developed.

  The older intradermal skin tests may be used but the antigen is difficult to obtain and standardization is a problem. Though quite popular at one time, it is not used nowadays.
- *Molecular diagnosis*: Various formats of the PCR may be used as well as the Loop-mediated isothermal amplification assay (LAMP). For PCR, the gene targets include Internal Transcribed Spacer Regions, the mitochondrial Cytochromec Oxidase Subunit 1, and Nicotinamide adenine dinucleotide-hydrogen (NAD Dehydrogenase Subunit 1). The Ty3/gypsy-like long-terminal repeat retrotransposon sequence has been used as a target in LAMP assay. Nested PCR can be used to distinguish between the species which is not possible in the existing LAMP assay. The molecular tests are useful if serodiagnostic tests are not available and for species differentiation.

FURTHER READINGS

Parija SC, Mahanta J, Chaudhury A, Devi KR, Ghosh T, Omar BJ. Guwahati declaration 2024: Indian Academy of Tropical Parasitology Guidelines for the diagnosis of endemic haemoptysis. *Trop Parasitol.* 2025;15:125–127.

Parija SC, Sehgal R, Ghoshal U, Khurana S, Khanna V, Chanda DD, Ghosh T, Bhosle NK. Pondicherry declaration on the identification and detection of *Entamoeba histolytica*. *Trop Parasitol.* 2020;10:71–73.

Index

Pages in *italics* refer to figures and pages in **bold** refer to tables

D

E

Q

R

S

T

V

W

Y

Z

For Product Safety Concerns and Information please contact our
EU representative GPSR@taylorandfrancis.com Taylor & Francis
Verlag GmbH, Kaufingerstraße 24, 80331 München, Germany